International Conference on High Speed Vessels 2020

London, United Kingdom
1 - 2 July 2020

ISBN: 978-1-7138-8298-5

Printed from e-media with permission by:

Curran Associates, Inc.
57 Morehouse Lane
Red Hook, NY 12571

Some format issues inherent in the e-media version may also appear in this print version.

The Royal Institution of Naval Architects
8-9 Northumberland Street
London, WC2N 5DA
United Kingdom

Phone: 020 7235 4622
Fax: 020 7259 5912

publications@rina.org.uk

Additional copies of this publication are available from:

Curran Associates, Inc.
57 Morehouse Lane
Red Hook, NY 12571 USA
Phone: 845-758-0400
Fax: 845-758-2633
Email: curran@proceedings.com
Web: www.proceedings.com

International Conference

High Speed Vessels

1st-2nd July 2020, London, UK

PAPERS

THE ROYAL INSTITUTION OF NAVAL ARCHITECTS
8-9 Northumberland Street, London, WC2N 5DA, Telephone: +44 (0)20 7235 4622

The Royal Institution of Naval Architects

INTERNATIONAL CONFERENCE

High Speed Vessels

1st – 2nd July 2020, London, UK

Papers published in these conference Proceedings are peer reviewed.

THE ROYAL INSTITUTION OF NAVAL ARCHITECTS
8-9 Northumberland Street
London WC2N 5DA

Telephone: 020 7235 4622
Fax: 020 7259 5912

CONTENTS

OPERATIONAL ASSESSMENT OF A FAST RESCUE CRAFT

M. Garvin and **T. Harris**, National Research Council, Canada

SUMMARY

Spark-ignited diesel engines were trialed aboard a Canadian Coast Guard fast rescue RHIB. The vessel was outfitted with a data acquisition suite to collect vessel and engine performance data and trialed in three seastates (approx. Beaufort 2 to 7). This paper focusses on the methodologies and results for several operational factors: ability to maintain course and speed, effort required to maintain course and speed, and impacts of seastate, course, and speed on fuel consumption.

The variability of steering input and course as a function of seastate, course relative to wave direction, and speed were used as a proxy for the difficulty and the effectiveness of maintaining course. The variability of throttle input and vessel speed as a function of sea state course relative to wave direction were used as a proxy for the difficulty and effectiveness of maintaining speed. The effect of seastate, course, and speed on fuel consumption were also examined.

NOMENCLATURE

Δ_{TEST}	Displacement as tested (kg)
δ	Steering angle of outboard motors (degrees)
φ	Throttle input, represented as engine rotational speed (degrees per second)
Ψ_{wave}	Course angle relative to wave direction (degrees)
B_{MAX}	Maximum Beam (m)
H_s	Significant Wave Height (m)
ka	Wave Steepness (dimensionless)
LCG_{TEST}	Longitudinal Center of Gravity, as tested, referenced from the intersection of the transom, keel, & centreline (m)
LOA	Length Overall (m)
T_p	Peak wave period (s)
$T_{P\delta}$	Peak period of steering input (s)
$T_{P\varphi}$	Peak period of engine rotational speed (s)
v_{boat}	Boat speed (m/s or knots)
v_{fuel}	Fuel volume (litres)
VCG_{TEST}	Vertical Center of Gravity, as tested, referenced from the intersection of the transom, keel, & centreline (m)
$\dot{x}$	Rate of change of x
$\bar{x}$	Mean value of x
σ_x	Standard deviation of x

1. INTRODUCTION

The work discussed herein developed from a series of test aboard a Canadian Coast Guard Fast Rescue Craft (FRC). The initial goal of the tests was to assess the performance of a set of new spark-ignited diesel outboard engines for use aboard the FRC fleet. Performance metrics of primary interest were: top speed, acceleration, towing capacity, reliability, and fuel consumption. The tests completed included a set of star pattern tests completed in three different wave states at five speeds which form the basis for this paper.

Facilitating the test were a group of career Search and Rescue FRC operators whose experience proved invaluable throughout. Conversations with these operators revealed a physiological and psychological toll from long-term exposure to Search and Rescue FRC operations. Absent specific human-factors data, the task of assessing physiological load from available data was undertaken. This paper involves an assessment of the relative degree of difficulty and success of maintaining course and speed in different wave states and at different speeds. Additional human-factors information such as Motion Sickness Index and Motion-Induced Interruption can be calculated from the data and may be assessed for a future paper.

2. MATERIALS

The test vessel was a Zodiac Hurricane 753 OB Rigid-Hill Inflatable Boat (RHIB) in Search and Rescue configuration with a pair of Mercury Marine DSI 3.0 spark-ignited diesel outboard motors, rated at 130 kW each. In this configuration, the helmsman's console is centerline, approximately midships with navigator consoles to port and starboard of centerline immediately aft of the helmsman [14]. All three consoles include saddle-style seats with shock absorbers. Throughout the test, three crewmembers (two SAR operators + one researcher) maintained the same nominal positions. The helmsman's controls include dual throttles and a steering wheel. The vessel has dual (forward and aft) fuel tanks which were filled prior to each test day. The vessel's principal particulars and pertinent details are [13]:

LOA = 8.15 m

B_{MAX} = 2.75 m

Δ_{TEST} = 2,837 kg

LCG_{TEST} = 2.32 m

VCG_{TEST} = 0.51 m

Fuel Cap. = 321 l (241 kg) aft, 181 l fwd. (136 kg)

Throughout the test series, the FRC was fitted with a data acquisition system which complimented the engines' on board telemetry (Mercury "SmartCraft"), providing a comprehensive record of the vessel's performance. During the star pattern tests, a wave buoy was deployed in the test area to capture wave state information. Sensors of interest to this work included (all sampled at 100 Hz): vessel speed and course (DGPS), steering angle of the outboard engines (LVDT between the engine and a fixed point on transom), engine rotational speed, and fuel consumption (both from "SmartCraft").

For the first two wave states tested, a TRYAXIS^{TM} Directional Wave Buoy [1] was moored near the test site. Since the third wave state was tested in deep water, mooring was not feasible so a TRYAXIS^{TM} Mini Wave Buoy [2] was connected to a drogue and allowed to drift downwind through the test area during the test.

3. METHODOLOGY

Testing was conducted in three sea states. Each test consisted of a series of straight legs at different courses relative to the wave direction. The target courses relative to wave direction ('0 deg' is head seas) for each test are shown schematically in Figure 1. The duration of each leg was adjusted based on the target boat speed, with slower speeds maintained for a longer duration. The star patterns were planned with target boat speeds of (leg duration in brackets): 4 knots (240 seconds), 8 knots (240 seconds), 20 knots (180 seconds), 30 knots (120 seconds), and full speed/40 knots (90 seconds). In head seas, this test program yielded at least 50 wave encounters in head seas for each speed [12]. It would have been preferable to use a longer leg duration to provide a larger number of wave encounters but the leg length could not be increased given the test speeds and the test area size limitations (approximately 1 nautical mile maximum leg length). Furthermore, there was noticeable crew fatigue with a full star test duration of approximately 2.5 hours during which time the crew experienced significant vibration and water spray.

In the first wave state, the 30 knot and 40 knot star tests were not completed due to concerns about the vessel becoming airborne in the short, steep waves; thus defining the upper edge of the vessel's speed performance envelope in these conditions. Indeed, the motions were so severe even at 20 knots that the engines' on board telemetry interpreted one wave impact as a collision and sent an automated message to the operators' supervisor that the vessel had been involved in a collision, resulting in some commotion and eventual laughter. In the third wave state, testing was only completed at 4 knots and 20 knots; higher speeds exceeded the vessel's speed performance envelope for the seastate and testing was cut short due to safety concerns.

Wave buoy data from the duration of the testing was analyzed using the wave buoys' on board software to calculate wave statistics: significant wave height (H_s), peak period (T_p), and steepness (ka) for each wave state tested [9]. The test wave states are summarized in Table 1.

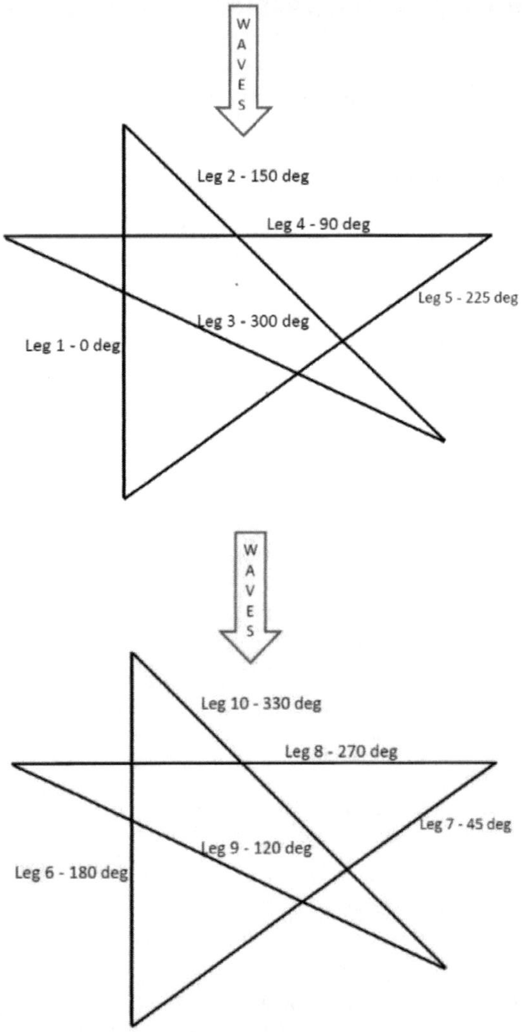

Figure 1. Star pattern test schematics, turning to Starboard (top) and to Port (bottom).

For wave state 1, the wind was approximately 20-25 knots, in line with the wave direction. Waves were primarily wind-driven with a short fetch of about 5 nautical miles. For wave state 2, the wind was less than 5 knots with waves entering the mouth of the bay where testing was conducted from open ocean. For wave state 3, the wind was approximately 15 knots, 30 to 50 degrees offset counter clockwise from the wave direction. Wave state 3 was tested near shore but in a stretch of water exposed to the open ocean. Current in the test areas is negligible.

Table 1. Summary of test wave states[a].

	H_s	T_p	ka
Wave State 1	0.52 m	2.71 s	0.67
Wave State 2	0.40 m	6.51 s	0.077
Wave State 3	3.6 m	11.9 s	0.52

[a] *For the full duration of the star test. For analysis, wave statistics were calculated for each leg of the star test.*

3.1 OPERATOR EFFORT & SUCCESS

Standard deviation and frequency of driver inputs (steering and throttle) have been used extensively in studies of automotive driver effort & performance. In many of the same studies, standard deviation of vehicle speed and lateral lane position are used to assess driver performance [6, 8, 10]. These techniques were adapted to assess FRC operator effort and success with throttle input, steering input, and speed being directly comparable metrics and course being analogous to lane position.

To investigate the operator's degree of success in maintaining course and speed, the standard deviation of course and speed from the mean achieved value was assessed for each leg to assess lateral/yaw control. The decision to assess the standard deviation from the mean rather than the difference between the achieved value and the target value was informed by the practicalities of navigating small craft in rough weather; delays in the instrument response made navigating by instruments impractical, causing the operators to result to alternative methods. For instance, operators would generally pick a distant visual waypoint and steer towards it rather than relying on their instruments. If this waypoint was not perfectly aligned with the target course, the course steered would systematically deviate from the target. This systematic variation was not a result of interest. In general, achieved boatspeeds were quite close to the target boatspeed but the same metric of deviation from mean achieved speed was used to assess longitudinal control.

The vessel was not fitted with a direct measurement of throttle input so engine rotational speed (averaged between port and starboard engines) was used as a proxy for throttle position. To assess the amount of effort required for the operator to maintain course and speed on each leg, the standard deviation of steering input and engine rotational speed from their respective means were calculated along with the peak frequency of this deviation. Specifically, the standard deviation of steering input and engine rotational speed for each leg give an indication of the magnitude of the required operator inputs while the peak frequency of the steering input and engine rotational speed changes gives an indication of how frequently a correction must be made. To calculate the peak frequency of the steering and engine rotational speed inputs, the following process was used:

1. A high-pass filter at 0.01 Hz cut-off was applied to the data,
2. the data was tared to its mean value,
3. an FFT was used to convert to frequency domain,
4. the frequency-domain data was smoothed using a 5-point moving average,
5. the peak frequency was calculated, and
6. the peak frequency [Hz] was converted to a peak period [s].

An example plot of the frequency domain data is shown in Figure 2. These plots were used to validate and interpret the calculated peak frequency values.

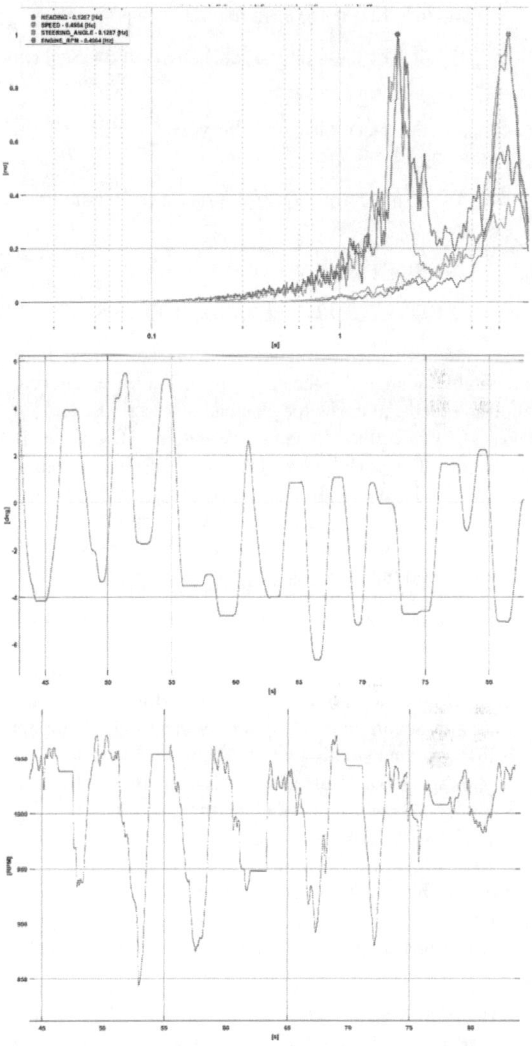

Figure 2. Example frequency-domain plot of throttle and steering input data (top), example of steering angle data trace (middle), and example of engine rotational speed data trace (bottom).

3.2 FUEL CONSUMPTION & RANGE

Fuel consumption as recorded by the engines' on board telemetry system was also averaged for each leg of the star tests. This allowed the effect of wave state, course relative to wave, and speed on fuel consumption to be assessed. To express the fuel consumption in operational terms, the estimated operational range (fuel reserve of 20%) was calculated for each leg, using the following equation: $\boldsymbol{Range_{Operational}} = \frac{v_{fuel} \times \bar{v}_{boat}}{\bar{v}_{fuel}}$

$$\boldsymbol{Range_{Operational}} = \frac{\boldsymbol{v_{fuel}} \times \boldsymbol{\bar{v}_{boat}}}{\boldsymbol{\bar{v}_{fuel}}}$$

where:

$Range_{Operational}$ is the distance the vessel can travel leaving 20% fuel reserve (in nautical miles),

v_{fuel} is 80% of the total fuel capacity of the vessel (in litres),

$\boldsymbol{\bar{v}_{boat}}$ is the mean speed of the vessel for the leg (in knots), and

$\boldsymbol{\bar{v}_{fuel}}$ is the mean fuel consumption for the leg (in litres per hour).

4. RESULTS

4.1 EFFORT & SUCCESS – STEERING

In wave state 1 (the shortest, steepest seastate tested) there is a notable decrease in the standard deviation of steering input as speed increases for all headings (σ_{δ} = 1.0 to 5.4 degrees), while the standard deviation of course is consistently low for all speeds (σ_{Ψ} = 1.9 to 9.3 degrees), as shown in Figure 3. This hints at increased directional stability with speed, reducing the magnitude of the steering corrections required to maintain course and was evident aboard the boat.

The largest standard deviations in course (σ_{Ψ} up to 29 degrees) were observed in starboard bow quartering to starboard beam waves in wave state 2 (the longest, gentlest seastate tested) at full speed, shown in Figure 4. These results are accompanied by increased steering inputs (albeit lower than steering inputs seen in wave state 1; σ_{δ} up to 3.0 degrees) and hint at some dynamic instability that was evident aboard the vessel in these conditions. It is unclear, however why this result is not symmetrical with port bow quartering to port beam seas showing far lower numbers. These results aside, steering inputs and the standard deviation of course are quite small, indicating that little effort is required to maintain a consistent course.

The dynamic stability noticed at speed is again evident in the steering inputs and standard deviation of course observed in wave state 3 (the largest wave state tested), shown in Figure 5, with the steering inputs and course standard deviations lower at 20 knots boatspeed than at 4 knots. Particularly in beam and following seas, quite large steering inputs (σ_{δ} up to 3.6 degrees at 4 knots and 2.0 degrees at 20 knots) are needed to maintain course (σ_{Ψ} = 9.2 to 14.6 degrees at 4 knots, 2.8 to 8.7 degrees at 20 knots). In the case of following seas, this appears to be related to a tendency for the vessel to "surf" down the face of the wave which was very apparent aboard the vessel, resulting in a marginal loss of control observable through the increased course standard deviation in following seas.

Peak steering input frequency data are quite consistent across all conditions (Figure 3, Figure 4, & Figure 5). Across all headings and wave states, median peak steering input period is 7.5 seconds with little variation (1st Quartile: 6.5 seconds, 3rd Quartile: 8.2 seconds).

4.2 EFFORT & SUCCESS – SPEED

In wave state 1, throttle inputs (using engine rotational speed as a proxy) are relatively consistent for all headings and speeds with slight increases in head and following seas and at higher speeds (σ_{φ} = 130 to 1,540 degrees/second), as shown in Figure 6. The standard deviation of speed, however, significantly increases with boatspeed in these waves ($\sigma_{v_{boat}}$= 0.21 to 2.3 knots). This is likely to the significant slamming felt at higher speeds in this condition. In head seas, there were several cases of propeller ventilation as the propeller came close to the water surface during extreme pitching motions. Overall, these were not frequent enough to have caused a significant increase in the standard deviation of engine rotational speed but they may be a contributing factor to the increased standard deviation in head seas where these events were typically observed.

In wave state 2, shown in Figure 7, standard deviation of engine rotational speed and the standard deviation of boatspeed are relatively small at all headings (σ_{φ} = 120 to 1,120 degrees/second, $\sigma_{v_{boat}}$ = 0.17 to 1.8 knots), slightly larger in head and following seas at speeds 20 knots and above. At the highest speed, there is also an increase in engine rotational speed and boatspeed standard deviation noticed in beam seas. This is potentially linked to the directional instability observed at this condition.

The largest standard deviation of engine rotational speed and boatspeed are observed in wave state 3, shown in Figure 8, at 20 knots and at all headings with a σ_{φ} of 1,230 to 3,100 degrees/second required to maintain a $\sigma_{v_{boat}}$ of 1.6 to 3.3 knots. This is expected to be linked to the operator making constant throttle adjustments to control the vessel; as the vessel approached the crest of the wave, the throttle would be reduced to prevent the vessel from becoming airborne and in troughs, the throttle would be increased to drive the vessel up the wave. In conversation with the crew, these adjustments are vital at all headings in large, steep, fast-moving waves to avoiding loss of control and capsize. At 4 knots in this wave state, less adjustment of engine rotational speed (σ_{φ} = 250 to 350 degrees/second) is required to maintain a very consistent boatspeed ($\sigma_{v_{boat}}$ = 0.4 to 1.1 knots).

The peak period of engine rotational speed is quite consistent (albeit less consistent than for steering input) across all conditions examined (Figure 6, Figure 7, & Figure 8) with a median value of 6.9 seconds (1st Quartile: 5.2 seconds, 3rd Quartile: 8.0 seconds). Interestingly, the peak period of engine rotational speed appears lower (indicating more frequent adjustments) at both 4 knots (median $T_{P\varphi}$ = 4.8 seconds across all wave states) and 30 knots (median $T_{P\varphi}$ = 2.7 seconds), particularly in beam seas where a minimum $T_{P\varphi}$ of 1.2 seconds is observed.

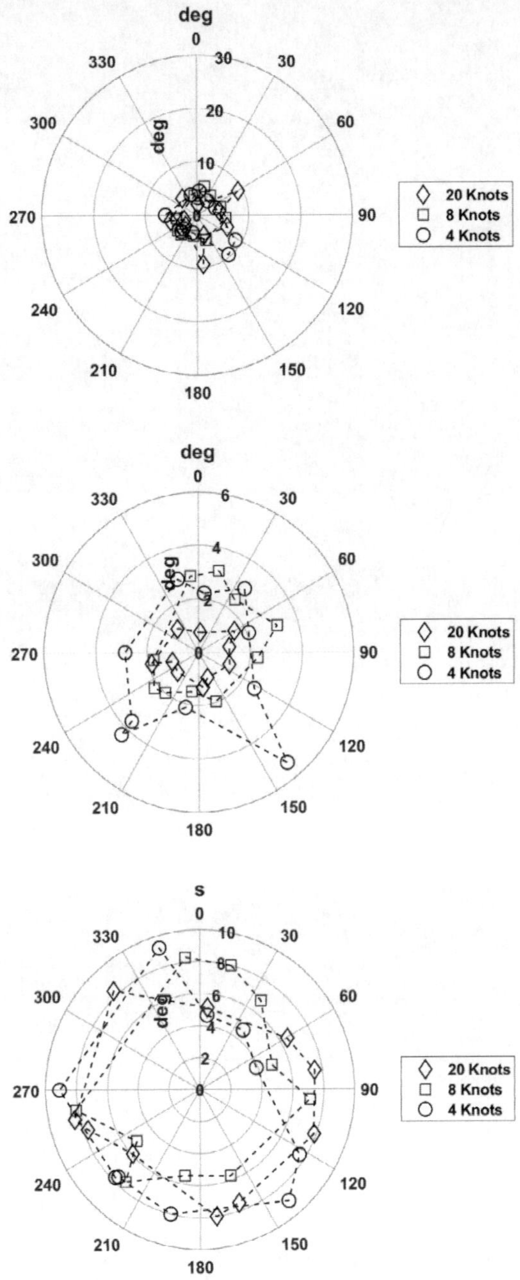

Figure 3. Standard deviation of course (top), standard deviation of steering input (middle), and peak frequency of steering input (bottom) for star test in Wave State 1 ($H_s = 0.52m$ and $T_p = 2.71s$).

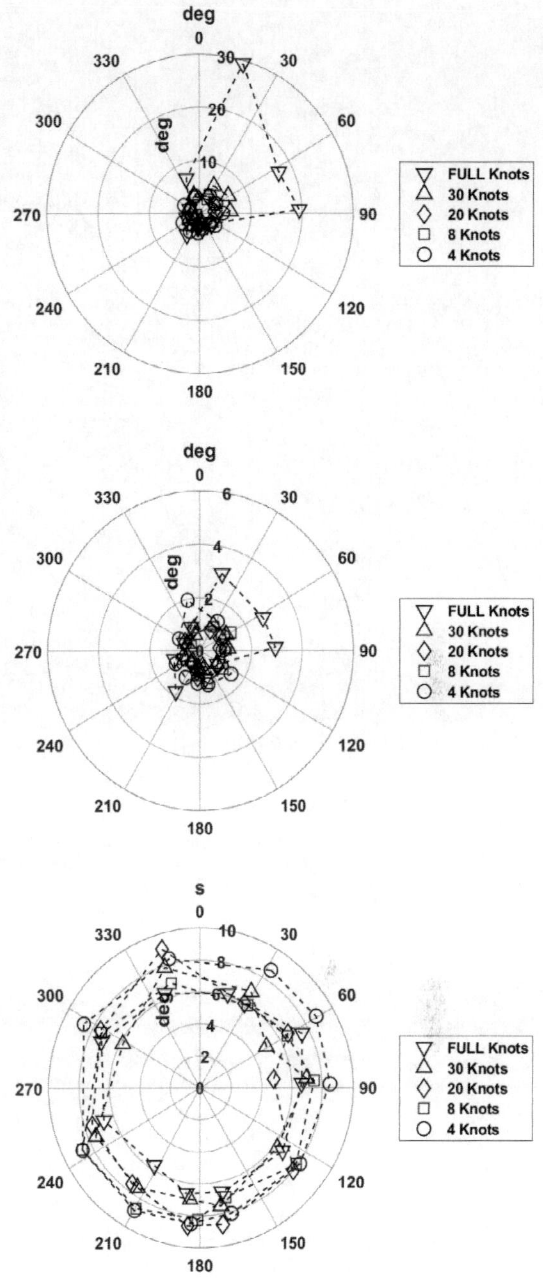

Figure 4. Standard deviation of course (top), standard deviation of steering input (middle), and peak frequency of steering input (bottom) for star test in Wave State 2 ($H_s = 0.40$m and $T_p = 6.51$s).

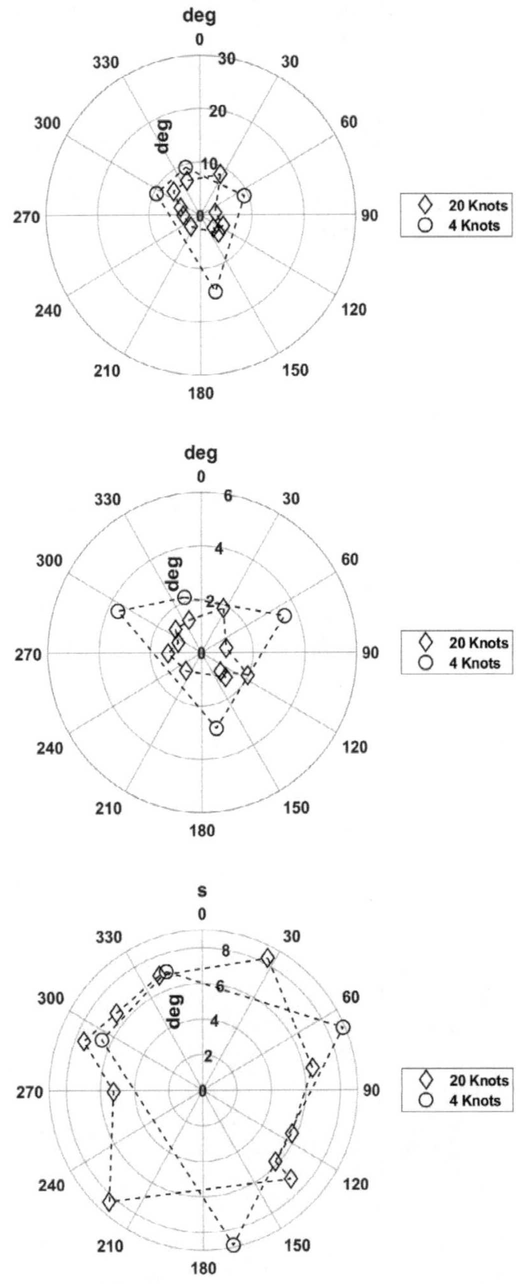

Figure 5. Standard deviation of course (top), standard deviation of steering input (middle), and peak frequency of steering input (bottom) for star test in Wave State 3 ($H_s = 3.6m$ and $T_p = 11.9s$).

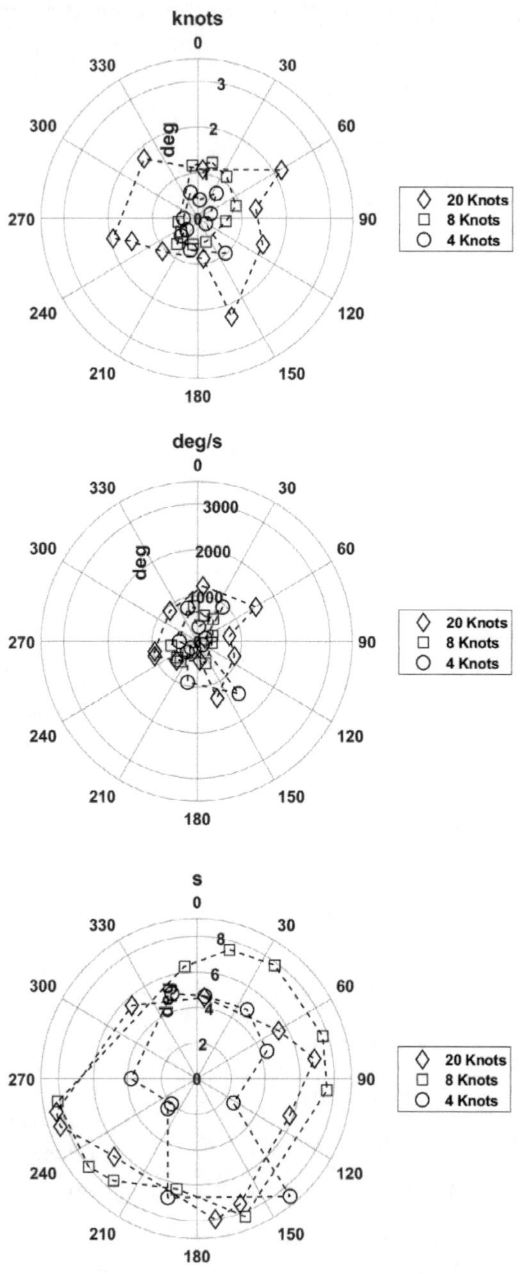

Figure 6. Standard deviation of boat speed (top), standard deviation of engine rotational speed (middle), and peak period of engine rotational speed (bottom) for star test in Wave State 1 ($H_s = 0.52m$ and $T_p = 2.71s$).

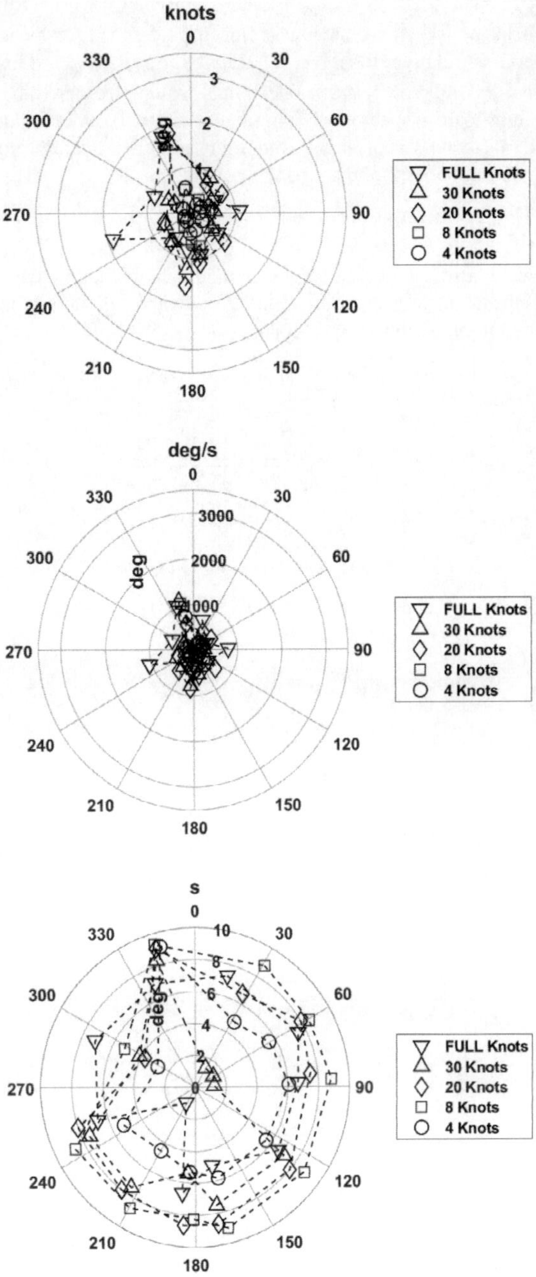

Figure 7. Standard deviation of boat speed (top), standard deviation of engine rotational speed (middle), and peak period of engine rotational speed (bottom) for star test in Wave State 2 ($H_s = 0.40m$ and $T_p = 6.51s$).

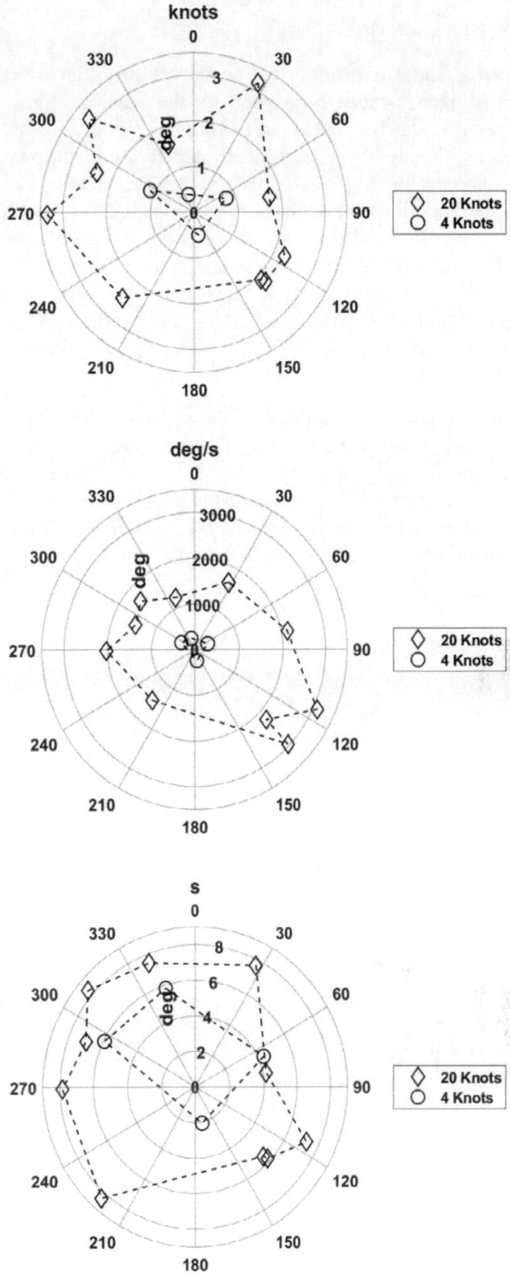

Figure 8. Standard deviation of boat speed (top), standard deviation of engine rotational speed (middle), and peak period of engine rotational speed (bottom) for star test in Wave State 3 ($H_s = 3.6m$ and $T_p = 11.9s$).

4.3 FUEL CONSUMPTION & RANGE RESULTS

Figure 9 shows the mean fuel consumption for each leg plotted against mean boat speed for the leg. This figure shows a fairly linear relationship between fuel consumption and boat speed, with each knot of additional boat speed requiring approximately 10 additional litres of fuel per hour throughout the operational range. It is worth noting that fuel consumption is a secondary consideration in the operational profile for this type of vessel. In responding to marine emergencies the prerogative is to reach the stricken vessel or search area as quickly as possible. Fuel consumption is a consideration only insofar as it limits the operational range of vessel.

Figure 10 shows an estimate of the operational range of the vessel, allowing for a fuel reserve of 20%. Disregarding operational speeds of below 10 knots as not fitting the operational profile typical of this vessel, a maximum range of approximately 50 nautical miles is achieved between 20 knots and 30 knots.

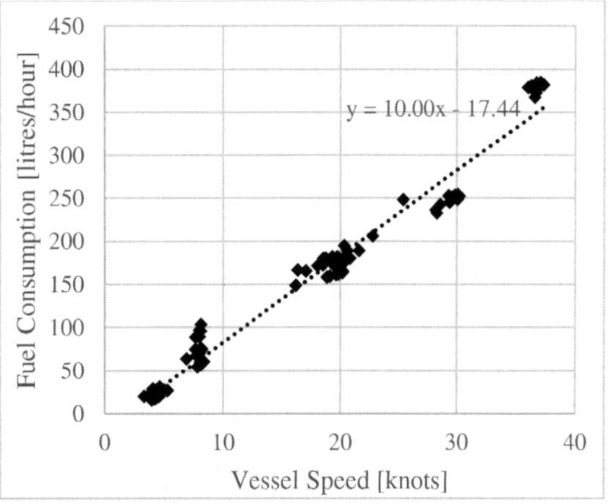

Figure 9. Fuel consumption (averaged for each leg) as a function of boat speed.

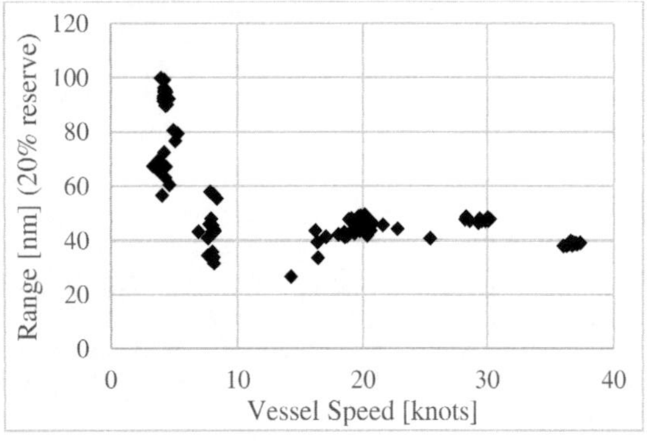

Figure 10. Predicted range [nautical miles] as a function of boat speed, allowing for a 20% fuel reserve.

4.4 CORRELATIONS

Correlations were calculated between steering input and engine rotational speed standard deviations & peak periods ($\sigma_\delta, \sigma_\varphi, T_{P\delta}, T_{P\varphi}$) and the significant wave height, peak period, wave steepness, mean boat speed, & mean course relative to wave ($H_s, T_p, ka, \bar{v}_{boat}, \bar{\Psi}_{wave}$). This was done to better understand which factors have significant impact on the amount of effort (size of input and frequency of input) required from the operator to maintain course and speed.

The correlations between course and speed standard deviations ($\sigma_\Psi, \sigma_{v_{boat}}$) and $H_s, T_p, ka, \bar{v}_{boat}, \& \bar{\Psi}_{wave}$ were also calculated to understand which factors have significant impact on the operator's degree of success in maintaining course and speed.

The equation used to calculate the correlation coefficients is shown below:

$$\rho(X,Y) = \frac{\sum(x-\bar{x})(y-\bar{y})}{\sqrt{\sum(x-\bar{x})^2 \sum(y-\bar{y})^2}}$$

where:

$\rho(X,Y)$ is the correlation coefficient between X and Y,

x represents $\sigma_\delta, T_{P\delta}, \sigma_\varphi, T_{P\varphi}, \sigma_\Psi, or\ \sigma_{v_{boat}}$ for each leg, and

y represents $H_s, T_p, ka, \bar{v}_{boat}$, or $\bar{\Psi}_{wave}$ for each leg.

A summary of correlation coefficients is found in Table 2.

Table 2. Correlation coefficients (moderate correlations in bold, strong correlations in bold italics).

Correlation between σ_δ (steering input) and:		Correlation between $T_{P\delta}$ (steering input) and:	
$\bar{\Psi}_{wave}$	-0.1200	$\bar{\Psi}_{wave}$	0.1716
H_s	0.1302	H_s	-0.0503
T_p	-0.4345	T_p	0.0896
$\bar{v}_{boat}$	-0.3883	$\bar{v}_{boat}$	-0.2653
ka	**0.5426**	ka	-0.1336

Correlation between σ_Ψ (course) and:	
$\bar{\Psi}_{wave}$	-0.2068
H_s	0.1813
T_p	0.0354
$\bar{v}_{boat}$	0.1642
ka	0.1080

Correlation between σ_φ (engine rot'l speed) and:		Correlation between $T_{P\varphi}$ (engine rot'l speed) and:	
$\bar{\Psi}_{wave}$	-0.0585	$\bar{\Psi}_{wave}$	0.0485
H_s	**0.6094**	H_s	-0.0290
T_p	0.1005	T_p	0.0519
$\bar{v}_{boat}$	0.1835	$\bar{v}_{boat}$	0.0130
ka	0.3440	ka	-0.0544

Correlation between $\sigma_{v_{boat}}$ and:	
$\bar{\Psi}_{wave}$	-0.0933
H_s	**0.5885**
T_p	0.1315
$\bar{v}_{boat}$	0.3612
ka	0.3060

Correlation between $\bar{v}_{fuel}$ and:	
$\bar{\Psi}_{wave}$	0.0508
H_s	-0.0143
T_p	0.3157
$\bar{v}_{boat}$	***0.9823***
ka	-0.3212

5. DISCUSSION

The amount of effort put into maintaining course is most strongly correlated to wave steepness followed by wave period and boat speed. The effort is generally greater in steep waves and at lower speeds, generally decreasing with speed until an inconsistent increase is noticed at full speed. Despite the greater effort, the success of maintaining course is only very weakly correlated to the factors investigated, indicating that the operators are able to closely follow a course in all of the tested conditions except for the inconsistent instability at the highest speeds discussed above. While dynamic stability general increases with increasing speed, it may decrease at higher speeds due to a combination of bow steering, chine riding, and porpoising although these phenomena are highly dependent on factors such as longitudinal center of gravity and engine trim [3] (which was left to the discretion and experience of the operator in our case). These phenomena are also highly dependent on the details of the hullform with changes in roll and directional stability achievable through modifying trim and deadrise angles [5, 11]. In our case, the potential for such instabilities is only observed at the highest speeds and then it is not consistently observed. Given the complex geometry of the inflatable tube atop the deep vee rigid hull, it is possible that a slight static roll or geometric asymmetry could result in the asymmetry of the observed response.

Wave height has the biggest influence on the operators' ability to maintain speed and the effort required to try and maintain speed. There is also a weak correlation with wave steepness. Conceptually, this may be due to the need to a tendency to accelerate while riding down a wave and decelerate as it rides up a wave, with this effect more pronounced in larger, steeper waves. In an effort to maintain speed, the operator must vary the throttle in an attempt to maintain speed. In operational conditions, there may be less emphasis on maintaining speed although operators spoke of a need to quickly accelerate and decelerate, particularly while heading into large and/or steep waves to prevent slamming and, in extreme cases, maintain control to avoid capsize or "kiting" whereby the boat becomes airborne off the crest of a wave. When discussing performance criteria for candidate engines, the crew ranked this ability to accelerate rapidly or "punch through waves" as the most important.

Examination of the raw data traces for steering and engine rotational speed yielded important insights: there was a tendency for the operators to make relatively rapid adjustments and then hold the input constant for a period of time before making an adjustment in the opposite direction (this is observable in Figure 2). Notionally, this is due to the operators noticing a deviation from the target, adjusting, and then holding the input constant while waiting for the vessel to respond. It is expected that this type of "adjust-hold-adjust" strategy dominated the frequency domain results. This may offer an explanation for the lack of clear correlation between any of the investigated parameters and the frequency of operator inputs; operator training/sensitivity and the vessel's rate of surge and yaw response dominate the other factors. This characteristic may be expected to be different for vessels driven by an autopilot. Across much of the data, engine rotational speed frequency distributions hint at a "double peak". The higher-frequency peak may be due to the operators making more frequent small adjustments but throttle position is not the only factor determining engine rotational speed as the outboard motors are prone to varying speed as propeller loading changes. The raw data traces for engine rotational speed show higher-frequency periodic variations (~1+ Hz) superimposed on the adjust-hold-adjust characteristic, observable in Figure 2. The adjust-hold-adjust component of the data is reliably attributable to the human operator's input but, for the higher frequency inputs, the human inputs and the changes in rotational speed cannot be reliably separated.

6. CONCLUSIONS

Adapting methods used to assess driver performance in the automotive industry (standard deviations of steering input, throttle input, course, and speed in our case) has promise as a useful method of assessing the degree of difficulty and success of operating ships and boats in different conditions. In addition to assessing the effect of different environmental conditions, these methods could be used to assess operator effectiveness (potentially useful as a training aid) or autopilot effectiveness (potentially useful as an autopilot design and tuning aid).

It is recommended that direct measurements be made of all parameters of interest in this type of work. The decision to rely on engine rotational speed as a proxy for throttle position meant that it was impossible to distinguish between high-frequency variations due to propeller loading and those due to the operator's intent. While the data indicates an adjust-hold-adjust throttle input with a period about 7 seconds is prevalent for this vessel across the conditions investigated, there is potential that the operators are making smaller, more frequent adjustments that cannot be separated from the rotational speed variation due to propeller loading. A direct throttle position measurement would clarify this issue.

Viewed alone, the standard deviation data is difficult to interpret. Calculating the correlation coefficients provided very useful insight into how the various parameters were related. Applied to a larger body of data, this method would allow more general conclusions to be drawn.

It is also difficult to interpret the underlying causes of the observed correlations and vessel responses. Some insight may be achieved by analysing the results herein together with some of the fundamental dynamics characteristics of the vessel, such as natural periods of roll, pitch, etc. and the motions of the vessel. This is beyond the scope of this study and the natural responses of the vessel were not measured.

There are many challenges associated with conducting sea trials in rough weather but these are compounded when working aboard small open boats. The researchers aboard the boat are subject to significant motions, vibration, noise, and frequent saltwater spray. Even simple tasks such as relaying course information to the helmsman are difficult and we adopted methods employed by the SAR crew to communicate (such as a tap on the left shoulder indicating "turn to port" while a tap on the right shoulder indicates "turn to starboard"). Taking effective notes is incredibly difficult; the only solution was a waterproof notebook and pencil but, even then, the writing was often barely legible. Having a detailed plan prior to leaving the dock is very important and minimizes writing and calculation required on the water.

From an operational point of view, while RHIBs provide a SAR option that allows rapid and flexible deployment, their achievable speed is significantly degraded in poor weather. This means that rapid response may be more easily achievable with other marine assets. Boatspeed is, however, the only factor considered that significantly impacts fuel consumption with fuel consumption only weakly or very weakly correlated to seastate and course; this means that the range of the vessel is not greatly impacted by adverse weather.

7. ACKNOWLEDGEMENTS

The authors would like to acknowledge the contribution of the Canadian Coast Guard FRC crew: Neil Peet, Brian Pye, Glen Saunders, and Steve Sheppard. Their enthusiastic professionalism and expertise were crucial to this work. We would also like to acknowledge National Research Council Technical Officer Derek Butler who developed the relationship with the SAR crew and oversaw the preparation of the test vessel and wave buoys.

8. REFERENCES

1. AXYS TECHNOLOGIES INC., 'TRIAXYS Directional Wave Buoy' *Online: https://axystechnologies.com/products/triaxys-directional-wave-buoy/,* Cited: February 1, 2020.

2. AXYS TECHNOLOGIES INC., 'TRIAXYS Mini Wave Buoy' *Online: https://axystechnologies.com/products/triaxys-mini-directional-wave-buoy/,* Cited: February 1, 2020.

3. BLOUT, D. L. and CODEGA, L. T., 'Dynamic Stability of Planing Boats', *Marine Technology, 29 (4-12),* 1992.

4. VAN DEN BOOM, H. J. J., MENNEN, G. G. J., and VERKUYL, J. B., 'Recommended Practice for Speed Trials', 2006.

5. BROWN, W. P. and KLOSINSKI, W. E., 'Directional Stability Tests of Two Prismatic Planing Hulls', *Davidson Laboratory*, 1994.

6. GREEN, P., CULLINANE, B., ZYLSTRA, B., and SMITH, D., 'Typical Values for Driving Performance with Emphasis on the Standard Deviation of Lane Position: a Summary of the Literature', *Technical Report: SAVE-IT, Task 3a,* 2004.

7. HARRIS, T., 'Marine Performance Evaluation of a Fast Rescue Craft', *Master of Engineering Thesis, Memorial University,* 2018.

8. MCLEAN, J. R. and HOFFMANN, E. R., 'Analysis of Drivers' Control Movements', *Human Factors 13(5) (407-418),* 1971.

9. MILES, M. D., MANSARD, E., VANDALL, T., and PHILLIPS, R., 'Triaxys Directional Wave Buoy', *Canadian Coastal Conference*, 2003.

10. PAPANTONIOU, P., PAPADIMITRIOU, E., and YANNIS, G., 'Review of driving performance parameters critical for distracted driving research', *Transportation Research Procedia 25 (1796-1805),* 2017.

11. PAYNE, P. R., 'Contributions to Planing Theory', *Ocean Engineering, 7th ed. Vol. 22 (699-729)*, 1995.

12. SEAKEEPING COMMITTEE, 28TH ITTC, 'Recommended Procedures and Guidelines: Seakeeping Experiments', *ITTC Quality System Manual, 7.5*, 2017.

13. ZODIAC HURRICANE TECHNOLOGIES, INC. 'SOLAS H753 Stability Report', *Zodiac Hurricane Product Data*, 2015.

14. ZODIAC HURRICANE TECHNOLOGIES, INC., 'H753 OB SAR, General Arrangement', *Zodiac Hurricane Product Data*, 2015.

9. AUTHORS BIOGRAPHY

Matthew Garvin holds the current position of Research Engineer / Harsh Environment Marine Operations Team Lead at the National Research Council's Ocean, Coastal, and River Engineering Research Centre in St. John's, NL, Canada. He is responsible for coordinating the research activities of the team with a focus on marine safety and foundational technologies to safeguard human life through automation and autonomy. His previous experience is mainly in private sector engineering consulting to clients in the marine and energy sectors, including an extensive list of test campaigns in the NRC's towing tank, ice tank, and offshore engineering wave basin.

Trevor Harris holds the position of Research Engineer at the National Research Council's Ocean, Coastal, and River Engineering Research Centre in St. John's, NL, Canada. Trevor's focus is on data analysis, particularly in handling ship performance datasets collected in the field and producing data products to help ship owners and operators to make informed operational decisions, usually with a focus on improving fuel economy. Prior to beginning work at NRC, Trevor worked in private sector engineering consulting as a data analyst supporting test campaigns in the NRC's towing tank, ice tank, and offshore engineering wave basin.

IMPROVING EFFICIENCY OF HIGH SPEED CRAFT USING MACHINE LEARNING AND BIG DATA

A. Rafiee and **M. Van-Someren** Austal, Australia

SUMMARY

In this paper we demonstrate how big data and state of the art machine learning techniques are used to build, test and deploy intelligent models to predict and optimize fuel consumption of High Speed Craft (HSC). This included addressing the challenges of predicting fuel consumption at high-speed in real-world conditions and providing real-time advice to the crew to optimize vessel settings to minimize fuel consumption.

1. INTRODUCTION

Fuel efficiency of ships has attracted much attention in recent years due to increasing environmental concerns and rising fuel costs. Societies increasing interest in environmental protection has led to the International Maritime Organization (IMO) taking measures towards limiting the environment footprint of ships significantly and reducing the total annual GHG emissions by at least 50% compared to 2008 by 2050 [8]. The Third IMO GHG Study in 2014 estimated that international shipping emitted 796 million tonnes of CO_2 in 2012, accounting for about 2.2% of the total global anthropogenic CO_2 emissions for that year, and that emissions from international shipping could grow between 50% and 250% by 2050 mainly due to the growth of the world maritime trade [17].

Trim optimisation is one of the easiest and cheapest methods for ship performance optimisation and fuel consumption reduction. It does not require any hull shape modification or engine upgrade and the optimisation can be done by proper ballasting at the start of the voyage and/ or adjustment of underwater control surfaces (if present) when underway.

Recent advances in data analytics have resulted in the increased popularity of data-driven models based on various Machine Learning (ML) techniques. The maritime industry, like other industries is also adopting ML based models for developing predictive models for various objectives such as fuel consumption, motion response and component failure. For example [10] used a "grey–box approach" to predict fuel consumption by combining feedforward neural Network and conventional physics based analysis. [16] used Artificial Neural Networks to predict fuel consumption of a product tanker and used the model to perform trim optimisations. [3] combined data driven ML models with physics based models to predict fuel consumption, and used that to optimise trim in real-world operations. It was shown that the data driven model was able to significantly improve the physics based estimations. [9] used ML data driven methods to study the relationship between vessel performance and various vessel and environmental variables. [1] used the Auto–Sklearn model of the Scikit–Learn package, a popular ML framework, to build fuel consumption prediction models for a cruise ship.

Since early 2017 Austal has recognized the value of smart technologies, Artificial Intelligence (AI) and the Internet of Things (IoT) in improving the performance and comfort of the HSC built by Austal. As evidence of this, Austal has developed the MARINELINK-Smart technology to transform Austal-built vessels into *Smart Ships* by gathering large volume of data from various sensors on-board. Austal has installed MARINELINK-Smart technology on multiple large HSC operating globally. Through the implementation of MARINELINK-Smart technology Austal has collected large operational data sets. These include vessel motions, vessel particulars, engine performance, motion control settings, as well as other relevant parameters.

In this paper, we demonstrate how big data and state of the art machine learning techniques are used to build, test and deploy intelligent models to predict and optimize fuel consumption of a high speed craft (HSC). We also show how the deployed models helped an operator to improve on their fuel consumption by up to 4.5%.

2. SHIP

The results described in this paper are for an Austal built high–speed vessel, Bocayna Express. Bocayna Express is operated by Fred. Olsen Express, one of the world's largest operators of large high-speed craft. In this section a brief introduction to the ship is provided.

Bocayna Express is a 66.2m Catamaran (see Figure 1). The vessel travels between the Canary islands of Fuerteventura and Lanzarote in the Atlantic Ocean. The journey is 8nm in length and scheduled to last 25 minutes [5].

Figure 1: Austal built, Fred Olsen's Bocayna Express.

Table 1 shows the general information of the vessel.

Table 1: Bocayna Express characteristics

Parameter	Value	Unit
Length overall	66.2	m
Length waterline	59	m
Beam	18.2	m
Draught	2.5	m
Speed	34	kn
Fr	0.72	–

3. DATA ACQUISITION

Table 2 shows the data gathered from sensors on-board of vessel.

The data is gathered automatically on–board of the vessel, timestamped and transferred for storage in Austal's secure cloud-hosted storage service.

In addition to vessel data, metocean conditions were also identified for each journey in the dataset. The metocean data was provided by TideTech [20]. Figure 2 shows an example of the data gathered on–board of Bocayna Express during a sailing from Correlajo to Playa Blanca.

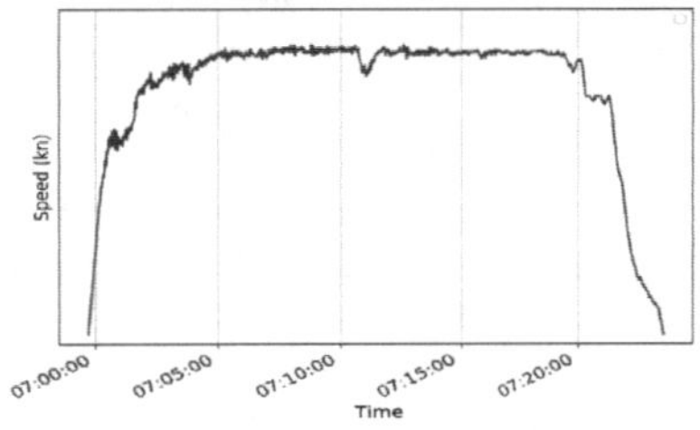

(a) Speed over Ground

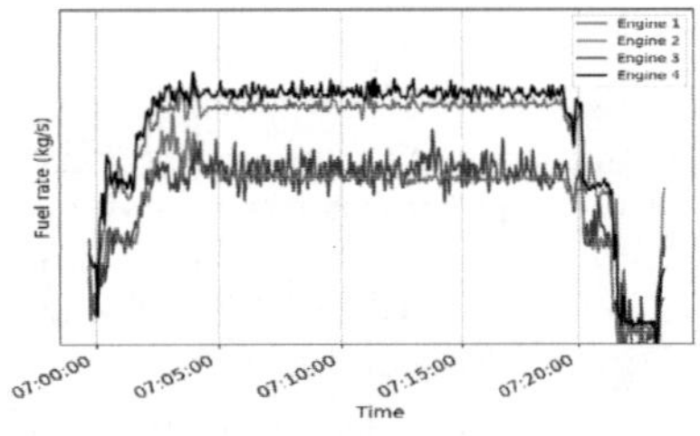

(b) Engine Fuel Rate

Figure 2: Sample of data collected during a journey from Bocayna Express.

4. PREDICTION OF FUEL CONSUMPTION

In order to build predictive models each journey was decomposed to segments of 3 minutes length. The duration of the segments was selected to ensure sufficient time to capture the influence of variations in parameters, whilst still being short enough to avoid strong smearing of these variations. Furthermore, segmentation of the journeys enabled the capture of the influence of water depth on vessel resistance (and hence fuel consumption) and vessel response (and consequently motion comfort).

As shown in Fig. 2a the vessel speed profile consists of three main sections, a ramp up zone when the vessel starts the journey and leaves port, a semi-constant zone when the vessel is cruising in open water and finally a ramp down zone as the vessel approaches the destination. To build the predictive models the ramp up and ramp down sections were removed from the data. Only the data during the cruising section where the vessel is travelling with semi-constant speed were considered.

4.1 DATA CLEANSING

Prior to build any predictive models it is necessary to clean the data and remove anomalies. This is done through the following steps:

I Time periods with missing data were removed from the datasets. It is possible to fill the gaps in data by interpolation but this would result in introducing errors in the final models.

II For each data column, the Z–score is estimated and then any data points with Z–score higher than 3 were removed from the datasets. Z–score of a data point is its relationship to the mean of the group of values, measured in terms of standard deviations from the mean. A Z–score of 3 means the data point is three standard deviations from the mean of data.

III The isolation forest [11] was used for the final anomaly detection and removals. The isolation forest efficiently uses random forests to perform outlier detection in high-dimensions. In this study the popular scikit-learn package [15] is used for isolation forest anomaly detection.

Once the data was cleaned various ML predictor models were developed. In the following sections, the details of the predictor models are explained.

4.2 PREDICTIVE MODEL

In contrast to physics driven models, the aim is to build a model for estimating fuel consumption from measured data. Given a dataset $D = \{(\mathbf{x}_1, \mathbf{y}_1), (\mathbf{x}_2, \mathbf{y}_2), \ldots, (\mathbf{x}_N, \mathbf{y}_N)\}$, where $\mathbf{x}_i \in \mathbf{X}$ is called the features and where $\mathbf{y}_i \in \mathbf{Y}$ are the labels, the objective is to build a mapping function $f : \mathbf{X} \rightarrow \mathbf{Y}$ as an approximation for the true function from the observed dataset. Generally, the approximate model can be built in three different approaches, namely, supervised learning, unsupervised learning and reinforcement learning [18]. There are a large number of supervised and unsupervised learning methods such as linear regression, lasso regression [19], kernel ridge regression [14], support vector machine (SVM) [4], gradient boosting [6], neural networks [13] and K-means clustering [12].

Here, the gradient boosting (GB) method is used to build the approximate function for fuel consumption from the measured data. GB is based on combining an ensemble of weak learners (prediction models) together to build a strong

Table 2: On-board data gathering.

Data	Frequency (Hz)	Description
Fuel flow meter	1	Fuel consumption rate of each engine
RPM	1	RPM of each engine
Ultrasounds	1	Distance to water surface at front and aft of the vessel
GPS	1	Latitude and longitude and heading of the vessel
SoG	1	Speed of vessel over ground
Ballast tanks	1	Levels of ballast tanks
Steering	1	Angles of waterjets
Interceptor/ Trim tab	1	Positions of interceptors or trim tabs on the vessel
Motion control surfaces	1	Positions of motion control foils
Motion control settings	1	Settings of the motion control system (*e.g* controller gain value)
IMU	5	Linear acceleration and rotational motion of the vessel

predictive model. This is achieved by sequentially building trees, aiming to reduce the loss function of the previous tree. Each new model is built up to reduce the residual of the previous model along the gradient direction.

Assuming an M stage GB model, at each stage m with a weak learner f_m, the approximate function can be improved by adding a new estimator $h_m(\mathbf{X})$ as:

$$h_m(\mathbf{X}) = \mathbf{Y} - f_m(\mathbf{X}) \quad (1)$$

Therefore, the GBM fits function h to the residual $\mathbf{Y} - f_m(\mathbf{X})$ and corrects the errors of the approximate function at the previous step.

In this paper, XGBoost library [2] is used for building predictive models using GB. XGBoost is an implementation of gradient boosted decision trees and designed for speed and performance.

4.3 RESULTS

To train the predictive model, the data was partitioned into training and testing datasets with a ratio of 80% and 20% of the total dataset, respectively. The XGboost was trained on the training set and the final model was tested against the test dataset. Furthermore, 10–fold cross validation is used for hyper parameters tuning of the XGboost model. Figures 3 and 4 show the accuracy of the optimum grey box predictive model on the training and testing datasets, respectively.

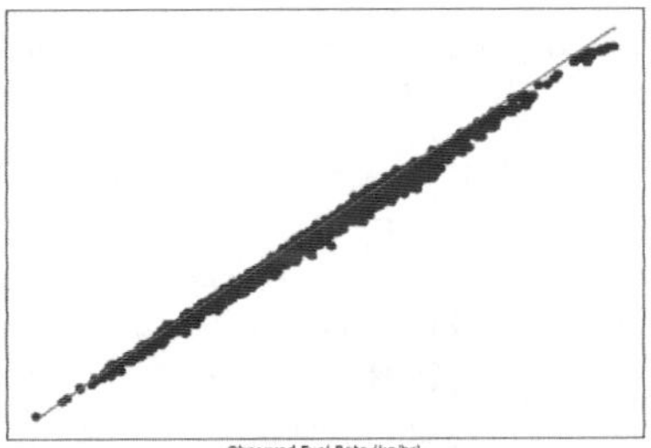

(a) Measured vs Predicted

Figure 3: Accuracy of the grey box fuel prediction model on the training set.

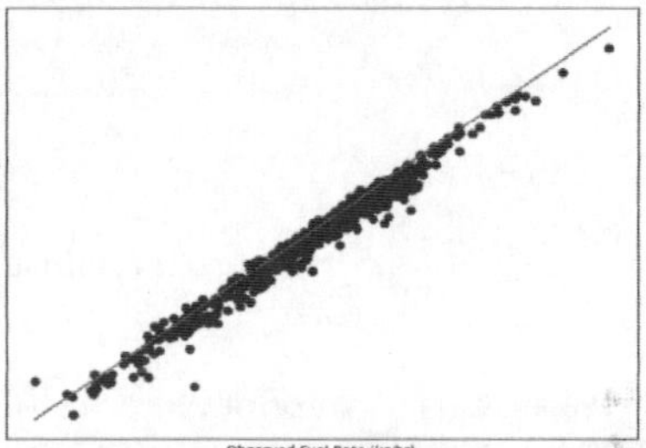

(a) Measured vs Predicted

Figure 4: Accuracy of the grey box fuel prediction model on the test set.

Finally, the importance of each feature on the final model is measured. The feature importance is a score showing how important the feature was to make key decisions with decision trees [7]. Figure 5 shows the feature importance for the predictive model. It can be observed that water depth, vessel speed, trim device position, vessel displacement and static trim also play important roles in fuel consumption prediction. This is of particular significance when the model is used for optimising vessel settings to minimize fuel consumption.

5. CONCLUSIONS

A data-driven predictive models were developed for predicting fuel consumption of a 65m high speed Catamaran operating in the Canary Islands. The predictive model was developed using XGboost [2], a gradient boosting framework. It was shown that the model can predict the fuel consumption with excellent accuracy when tested on a dataset that was not used during the training stage.

6. ACKNOWLEDGEMENTS

We thank Fred. Olsen Express for their ongoing collaboration with Austal in the development of the MARINELINK-Smart system.

We also thank the University of Western Australia (UWA) for their collaborative efforts in establishing the Austal

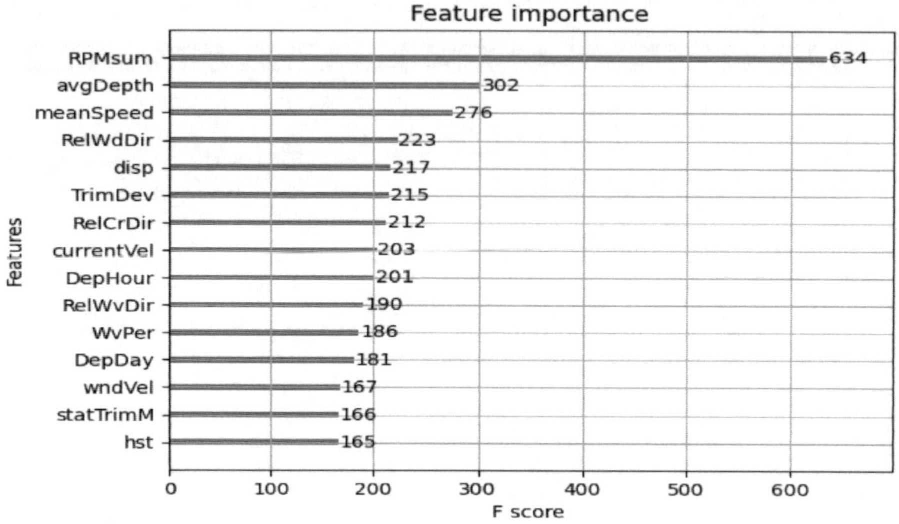

Figure 5: Feature importance of grey box fuel prediction model.

Smart Ship Initiative. The team developing MARINELINK-Smart is located at the UWA campus, where the ability to exchange ideas with students and staff on Machine Learning and other topics has been invaluable.

Finally we thanks Tidetech Pty Ltd, who have provided the metocean data which is a critical input to the analysis described in this paper.

References

[1] F. Ahlgren and M. Thern. Auto machine learning for predicting ship fuel consumption. In *Proceedings of ECOS 2018 - The* 31st *International Conference on efficiency, cost, optimization, simulation and enviromntal impact of energy systems*, 2018.

[2] T. Chen and C. Guestrin. XGBoost: A Scalable Tree Boosting System. In *Proceedings of the* 22nd *ACM SIGKDD International Conference on Knowledge Discovery and Data Mining*, pages 785–794. ACM, 2016.

[3] A. Coraddu, L. Oneto, F. Baldi, and D. Anguita. Vessels fuel consumption forecast and trim optimisation: A data analytics perspective. *Ocean Engineering*, (130):351–370, 2017.

[4] C. Cortes and V. Vapnik. support–vector networks. *Machine Learning*, (20):273–297, 1995.

[5] Fred Olsen, Bocayna Express.

[6] J. Friedman. Stochastic gradient boosting. *Computational Statistics & Data Analysis*, 38:367–378, 2002.

[7] T. Hastie, R. Tibshirani, and J. Friedman. *The Elements of Statistical Learning: Data Mining, Inference, and Prediction*. Springer, 2 edition, 2009.

[8] IMO. Adoption of the initial IMO strategy on reduction of GHG emmissions from ships and existing IMO activity related to reducing GHG emissiobs in the shipping sector, 2018.

[9] C. Jonge. Data-driven analysis of vessel performance, 2017.

[10] L. Leifsson, H. Saevarsdottir, S. Sigurdsson, and A. Vesteinsson. Grey–box modeling of an ocean vessel for operational optimization. *Simulation Modelling Practice and Theory*, 16:923–932, 2008.

[11] F. Liu, K. Ting, and Z. Zhou. Isolation forest. In *ICDM '08: Proceedings of the 2008 Eighth IEEE International Conference on Data Mining*, 2008.

[12] S. LLoyd. Least squares quantization in PCM. *IEEE Transactions on Information Theory*, IT–28(2):129–137, 1982.

[13] W. S. McCulloch and W. Pitts. A logical calculus of the ideas immanent in nervous activity. *The bulletin of mathematical biophysics*, 5(4):115–133, 1943.

[14] K. P. Murphy. *Machine Learning: A Probabilistic Perspective*. The MIT Press, 2012.

[15] F. Pedregosa, G. Varoquaux, A. Gramfort, V. Michel, B. Thirion, O. Grisel, M. Blondel, P. Prettenhofer, R. Weiss, V. Dubourg, J. Vanderplas, A. Passos, D. Cournapeau, M. Brucher, M. Perrot, and E. Duchesnay. Scikit-learn: Machine learning in Python. *Journal of Machine Learning Research*, 12:2825–2830, 2011.

[16] J. Petersen, O. Winther, and D. Jacobsen. A machine–learning approach to predict main energy consumption under realistic operational conditions. *Ship Technology Research*, 59:64–72, 2012.

[17] T. Smith, J. Jalkanen, B. Anderson, J. Corbett, J. Faber, S. Hanayama, E. O'Keeffe, S. Parker, L. Johansson, L. Aldous, C. Raucci, M. Traut, S. Ettinger, D. Nelissen, D. Lee, S. Ng, A. Agrawal, J. Winebrake, M. Hoen, S. Chesworth, and A. Pandey. *Third IMO Greenhouse Gas Study 2014*. International Maritime Organization, 2015.

[18] R. S. Sutton and A. G. Barto. *Introduction to Reinforcement Learning*. The MIT Press, 1998.

[19] R. Tibshirani. Regression shrinkage and selection via the lasso. *Journal of the Royal Statistical Society. Series B*, (58):267–288, 1996.

[20] Tidetech, 2020.

LAB TESTING OF SUSPENSION SEATS FOR HIGH-SPEED BOATS

Lab testing cannot determine seats' capacity to protect users from impact induced injuries.

J. Ullman, High Speed Boat Operations Pro, Gothenburg, Sweden
S.Å. Eriksson, Research Electronics AB Siljansnäs , Sweden
J.I. Kåsin, Norwegian Institute of Aviation Medicine, Norway
A. Borgen, Applica Test and Certification AS, fmr DNV Test Centre, Norway
D. Reynolds, University of Nevada Las Vegas, Mechanical Engineering. & Environmental Systems, USA

SUMMARY

Impact exposure onboard high-speed boats cause severe injuries and even lethal accidents. Suspension seats may reduce risks of injury. Real life testing of seats' capacity to protect users from injury is difficult to perform. Hence a functioning lab testing procedure would be of great value. Advertising and marketing campaigns claim that an ISO standard for lab testing suspension seats exists and that, by using this test method, a specific brand of seats has been scientifically proven to be the safest in the world. This prompted a scientific evaluation of the claimed standard, the relevance of the proposed test method and validity of its results. The study concluded that the method, claimed to be an ISO-standard, was not based on any scientific evidence and cannot test protective capacity of suspension seats and therefore may increase risks of injury. After these findings were presented to the ISO TC108 and its subcommittee SC4, the project CD 19470, was abandoned and the work group WG18 was dissolved. No functioning method for lab-testing of suspension seats' protective capacity has yet been proposed.

1. INTRODUCTION

Slamming-induced impacts onboard high-speed boats can cause severe injuries, lesions, permanent disabilities and sometimes even fatal overboard ejections. These problems seem to increase in numbers and severity along with an increasing number of increasingly faster boats coming in professional use. The problems seem to increase in spite of both standards and legislation for defining limits for supposedly dangerous exposure, as well as commercially available means developed to reduce the dangerous exposure.

Therefore, test standards and performance claims regarding protective capacity of protection equipment and protective methods must be based on solid evidence-based science.

The need for this study became apparent when government agencies were led to form their standards for suspension seats, based on claims that one specific kind of seats had been scientifically proven by tests made according to a new "ISO standard", to provide better injury protection than any other "on the planet".

1.1 PURPOSE OF THE STUDY

The primary objective of the study was to validate the test method, proposed to become an ISO standard, its relevance for tested seats' protective capacity, the validity and the repeatability of test results and the scientific validity of the claims and assumptions on which it was based.

A second objective was trying to identify what actual purpose the proposed test standard was intended for and how it was supposed to be used.

A third objective became to identify an alternative purpose that the proposed test standard could serve.

2.METHOD

2.1 EVALUATION OF A PROSED STANDARD

To get access to the document proposing a new ISO test-standard, the ISO Committee Draft CD 19470, it was necessary to join the ISO work group formed for the purpose, WG18, and the ISO Subcommittee, SC4, controlling it under the ISO Technical Committee, ISO TC108.

This was possible to do only by being appointed by the various national standardisation bodies as experts, representing the respective countries in the ISO committee. This gave full access to the documents and the WG meetings and the opportunity to commence the work in order to:

- Fact-check the claims founding the basis for the proposed method and standard.
- Understand the claimed purpose of the standard as outlined in the CD 19470
- Understand the real purpose of proposing a standard.
- Validate the assumptions founding the basis for the proposed method and standard.
- Validate the proposed use of the test results.
- Validate the relevance of test results for the claimed purpose.
- Validate repeatability of test results
- Validate the usability for certification by accredited labs

2.2 SUMMARY OF THE STANDARD PROPOSED IN ISO COMMITTEE DRAFT CD 19470

Seats should be tested for impact protection / shock-mitigating capacity by being attached to a solid platform, fitted with a downward pointing wedge on its underside. The platform with the seat should be

dropped from various heights into a sandbox or onto other resilient media NUD, to create peak level impacts at 4g, 7g and 10g on platform.
Incoming impacts should be measured with an accelerometer fixed to the platform. Impacts on the seat should be measured with an accelerometer placed on top of the seat padding.
On top of this dead weights of 40kg, 70kg and 100kg should be placed, representing various body weights
Seats should be tested only for purely vertical impacts, without any lateral components.
Seats capacity to protect humans from impact-induced injury should be calculated as the difference between the impact measured on the platform and the impact measured on top of the foam.
Before the comparison, impact data should be converted from pure acceleration levels to mean values of vibration using advanced algorithms and presented as VDV, Vibration Dose Value, Sed8, Seated Equivalent Dose for 8 hours, or as SRS, shock response spectrum.
The shape of the incoming impact on the platform was defined as a half sinus wave, which would fit into a predefined envelope which would specify aspects as rise-time, duration and peak impact level. Test impulses should simulate a half sinus wave of 100 ms duration.
Seats were to be assessed for protective performance by calculating the quote between the acceleration measured on the platform and the corresponding values measured on top of the seat cushion after these data have been processed into vibration does values.
Acceleration levels should be measured on the base platform, and on the top of the seat padding. Seats should be preloaded with dead weights, placed on top of the seat padding. Before being analysed and compared, registered acceleration data from discrete impacts should be low-pass filtered and processed by certain algorithms developed to quantify human exposure over time to vibration.
Seat manufacturers should do the testing themselves and present the results to the market. The presented results should help customers and end users make sure that the seats they buy would provide the level of impact reduction needed to protect users from injury.

3. THE BASIS OF THE PROSED STANDARD

The 12 cornerstones forming the base of the proposal were analysed along with the claims and assumptions by which each one was justified/defended.

3.1 "SEATS SHOULD BE TESTED ONLY FOR IMPACT LEVELS UP TO 10G"

Justifying claims: *Typical slamming induced impacts on hulls can go only up to 10 g and impacts above 10 g are "a one in million years event".*

Seats need to be tested only for typical and not for worst-case impacts.
These claims are false. Numerous user agencies have documented hull impact exposure levels exceeding 20g. See graph 1&2.
Impacts exceeding 10g are less common - but more dangerous than the less severe "typical" impacts. Seats approved for use by such lab testing, but not tested for real-life exposure levels may give a false feeling of security and hence cause increased risks of injury.
Justifying claim*: The test method is not intended to assess the seat response to the full range of conditions that may be encountered in practice, just to "typical impacts."*
This claim raises questions about the purpose of the testing. Suspension seats are normally acquired for the purpose of preventing impact induced injuries.
It is obvious that impacts with peak values between 10g and 20 g are more dangerous than those with peak values below 10 g.
The value of a standard for testing seats' protective function, only in the lower part of the actual exposure spectrum, was not explained.
The introduction of the CD clearly indicates that the purpose of the method is to test protective capacity. This can lead to a false feeling of safety and hence could increase risks of injury.

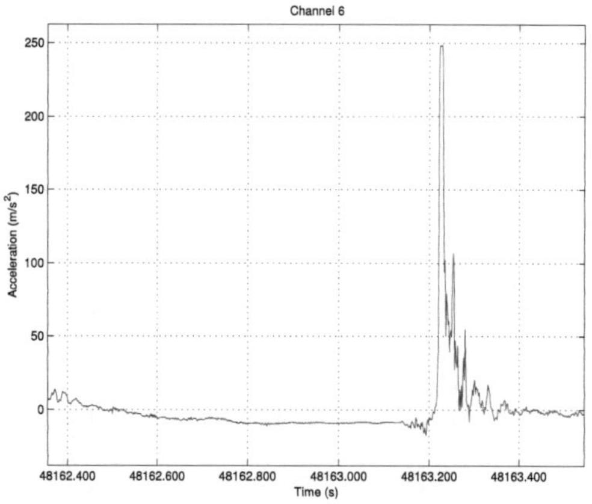

Graph 1. shows an impact measured on the hull of a Swedish 8m Coastguard RIB crossing the wake of a towboat at 23.7 kts. The pre-impact airtime, = freefall, lasts around 650 ms. The duration of the impact is ≈ 100 ms, and the rise time from 0g to peak level is 5 ms. This shape with, a steep rise time and a higher initial peak before a slow slant off, is characteristic of slamming impacts over the whole range of amplitudes.

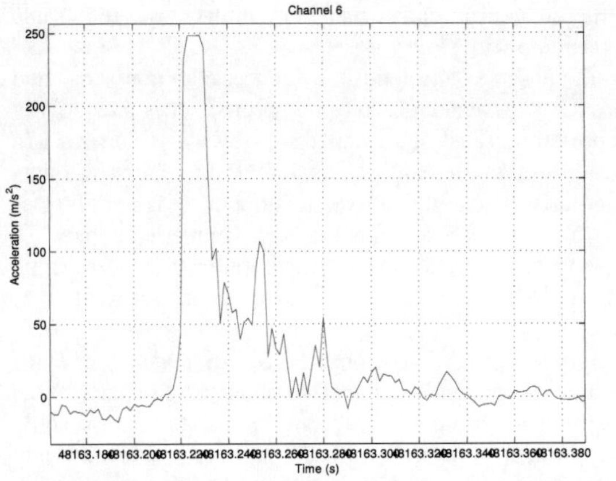

Graph 2. shows the same event, zoomed in on the x-axis for better visual analysis of the character of the impact. The cut-off shape of the curve shows that the impact exceeded the +25g range of the accelerometer.

It is well known that in road accidents, injury risks increase with the square of the speed at impact and that the kinetic energy is proportional to the square of the speed at impact. It is fair to assume that forces causing slamming induced injuries increase with the square of the speed at impact against the surface, which correlates to the measure impact values. Based on this assumption the relevance of testing seats' performance only to 10 g is questionable.

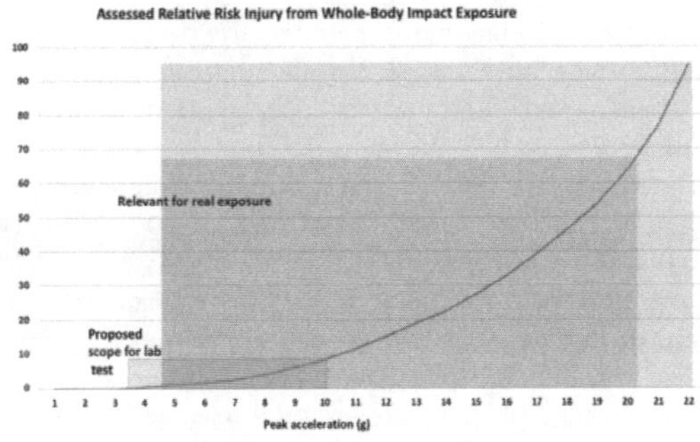

Graph 3. Shows an estimated relation between peak acceleration values and risk of failure in anatomical structures. This is not based on any experimental data and shown only for the purpose of illustrating the hypothesis that the injurie rise exponentially and linearly with the forces generated by the slamming impacts. This hypothesis can be tested in a prospective longitudinal exposure study designed for this purpose.i

3.2 "SEATS SHOULD BE TESTED ONLY FOR PURELY VERTICAL IMPACTS AND NOT FOR OBLIQUE IMPACTS CONTAINING EVEN LATERAL FORCES"

Justifying claims: "*Lateral impact forces are insignificant. Most impacts at sea are vertical. 81% of measured impacts are mainly vertical.*"

"Hence protection for lateral impact forces is not needed and seats capacity to mitigate later forces does not need to be tested"[ii].

This claim is false. It is well established that the human spine is more vulnerable to impact forces in the horizontal plane than it is to purely vertical.
The reasons for this are that exposure to impacts with significant horizontal components causes shear forces, bending forces, tear forces and rotational forces and that stabilizing and supporting structures of the spine are more effective in protecting it from the purely vertical.
Ii some cases, lateral forces on the hull can even exceed the purely vertical forces.[iii]
It is also well known that lateral impact forces acting on the human spine are considered more dangerous than the purely axial. It is common to evaluate lateral forces acting on the spine as causing an injury risk factor of 1.4 compared to pure axial forces.[iv]
Lateral forces also seem to play a significant role or in or be a prerequisite for overboard ejections.
Several people have lost their lives by being injured, due to overboard injections. If only 81% of impacts are mainly vertical, ≈ 1 out of 5 impact produce higher horizontal forces than vertical.
Structural failures seen in spinal injuries caused by slamming impacts at sea, tends to be asymmetrical. This indicates that even deflection of the spine is part of the etiology. Even the head jolt normally seen in severe impacts may be asymmetrical. [9]

3.3 "TEST IMPULSES FOR LAB TESTING OF SEAT IMPACT MITIGATION CAPACITY SHOULD BE IMPLEMENTED AS A HALF SINUS WAVES WITH 100MS DURATION"

Justifying claim: *A 100ms half sinus waved does simulate an impact measured on the hull of a high-speed boat hull slam. (Slam = impact with the water surface).*

This claim lacks scientific support. A 100ms half sinus wave is not representative of and does not simulate a slamming induced impact as measured at the hull of a high-speed boat. These do not even remotely resemble a half sine wave.
A 100ms half sinus wave has a risetime of 50ms, whereas real slamming impacts can have typical risetimes shorter than 10ms See Graph 5. A half sine wave misrepresents severely the actual rise time and the speed of onset.
With very hard low pass filtering of the actual boat slam impulse it can be converted to slightly resemble a half sine, and the shortest possible rise time that can be shown is that of the filter. This filtering can significantly reduce the actual peak values and increase the actual rise times.
This means that the test impulses used are not realistic and not relevant.
In fact a well-functioning seat suspension is for all practical matters a mechanical low pass filter. To test its

function on a signal already low pass filter is not scientifically acceptable and not relevant.
In this context it worth noting that even the human body functions as a low pass filter absorbing significant parts of the higher frequency content in the impacts. Soft tissue in the buttocks and in the and every segment of the spine serve as shock absorbers transforming impact energy from higher frequencies to lower frequencies. The most effective absorption in the body occurs when structures break, e.g. when vertebral bodies collapse.
There is, however, noting published indicating that even this process would convert the character of the impact to a half sine wave.

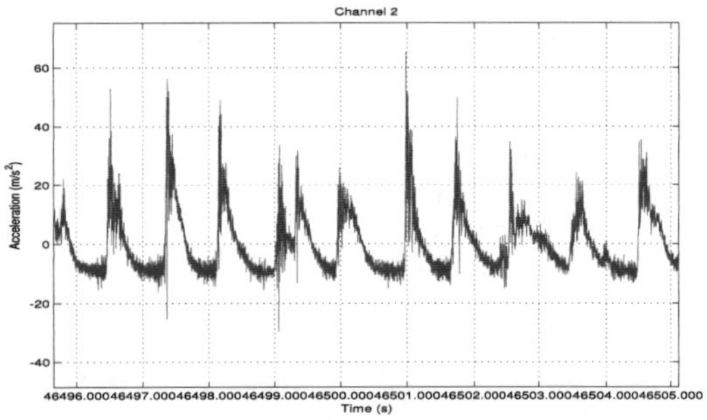

Graph 4 Showing moderate slams of 2-6 g with the typical shape: Short rise time and slow slant off; - not similar to any kind of half sine wave. The slamming is recorded in a sea state 2 on 14 m craft travelling at 55 kts. In this specific case it can noticed that the craft is airborne from impact to impact to impact with the hull is in free fall when it is not slamming.

3.4. "IMPACT DATA SHOULD BE LOW-PASS FILTERED AT 20 Hz BEFORE ANALYSES, BOTH THOSE MEASURED ON INCOMING IMPACTS AND THOSE MEASURED ON THE SEATS".

Justifying claims:

1. *"Components of impacts exceeding a frequency of 20Hz do not affect the human body".*
2. *"Energy content with frequencies in the 5-8 Hz spectrum is the most dangerous."*
3. *"Therefore all energy content in the frequency spectrum above 20Hz can be disregarded and filtered out."*

This L-P filtering is a severe methodological liability. It destroys data and hides relevant information about the real amplitudes, shown as peak values.
It produces false low peak values, which indicate false performance values.
It also hides all information about impulse risetime. The low pass filtering shows false low peak values, false longer rise times, and false low content of higher frequencies in the impacts. No support has been found in scientific literature, justifying the hiding of this information.
No support has been found in scientific literature for the claim that the human spine is not susceptible to the part of the energy of an impact which is above 20 Hz.
There are good reasons to believe that an impact with faster onset, shorter risetime, affects anatomical structures differently than an impact of the same amplitude with slower onset.
In the absence of scientific evidence, common sense and proven experience prevail must prevail. It is anecdotally reported that fast onset impacts, as by wave slams, of 7g can cause immediate pain, while fighter aircraft pilots regularly expose themselves to slow onset loads of 7g or more, without experiencing any immediate pain or injury. This may lead to the assumption that the risetime of whole-body impacts is a relevant factor for the risk of injury.
This needs new scientific methods to investigate. This matter is currently subject to analysis as part of an ongoing longitudinal prospective non-intervention study on human impact exposure at sea. Without any experimental data available to support either claim, that risetime is relevant or irrelevant, omitting or destroying potentially useful data is unacceptable.

Further claim justifying filtering use of vibration values:
4. *"The spine is most susceptible to injury by the energy content in discrete impacts around 5-8 Hz "*

This claim is not supported by scientific evidence.
It is most probably stemming from observations that cadaver specimens of lumbar spines seem to show an inherent resonance tendency around these frequencies, when exposed to continuous cyclic sinusoidal, perfectly axial, compressions and that such exposure over time can cause structural fatigue and eventually structural failure of endplates.
It is obvious that the experimentally found resonance frequency also depends on the size of the cadaver specimen and of the number of motion segments it contains, as well as the shape of spine specimen, neutral of flexed as it may be under impact, and also interindividual variations in body size and preload by body armour and helmets etc.
Most important however, is that there is no reason to believe that singular impacts cause any resonance at all, especially none that would keep resonating after a second, when the next impact could be expected to occur. Periods between relevant slamming induced impacts can for obvious reasons not be much shorter than a full second.
Video footage of humans, being exposed to severe impacts at sea, show that impacts are not perfectly linear and that the spine does not remain perfectly aligned but normally deflects.v
There is also no reason to believe that the resonance frequency provoked by sinusoidal compression of a perfectly aligned specimen of a lumbar spine, with five motion segments has any relevance for the biomechanical properties of a full-length spine, in vivo, when exposed to singular stochastic impacts from various directions.
Therefore, experimentally shown resonance frequencies of the human spine lack relevance for risk of injury caused by discrete impacts. Low pass filtering of measured data ruins much of the potentially most

relevant information in the gathered data especially, in the higher level- and faster onset impacts.

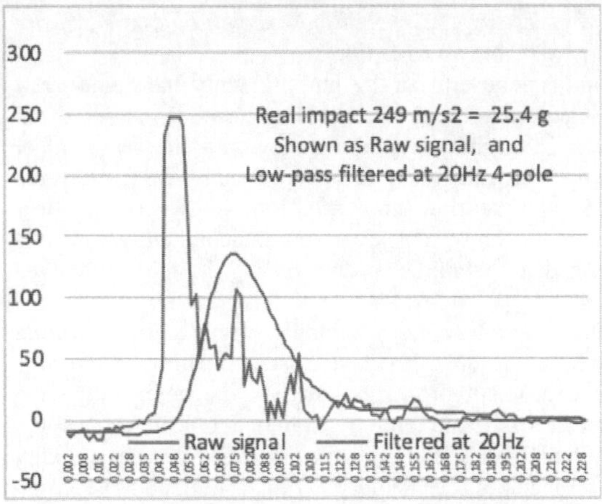

Graph 5. Shows a 25g hull impact, recorded onboard an 8 m Coastguard RIB at 26 knots. Comparing the recorded raw data to the same data low-pass filtered at 20 Hz with a 4-pole filter illustrates that both peak value and the rise time are severely changed. The peak value is reduced by more the 40% and the rise time increased 5 times, from 4 ms to 21 ms, reducing the jerk (j) by 90%, from ≈ 62000 m/s^3 to ≈ 6200 m/s^3.

3.5 "IMPACT DATA SHOULD BE REPORTED NOT AS THE MEASURED PEAK VALUES, BUT INSTEAD BE CONVERTED AND SHOWN AS SRS, VDV, OR SED8"

Justifying claim: *"SRS, shock response spectrum has been used to calculate transfer of impact induced forces in fighter aircraft ejection seats. VDV, vibration those value, has been used to quantify whole body vibration and is also the unit used in the Vibration-Comfort standard ISO 2631vi, and Sed8, seated equivalent dose for 8 hours, is used to quantify exposure to whole body vibration in an the updated ISO 2631:5".*

This conversion of data is a serious methodological flaw. None of these conversions will make the data possible to use to quantify the dangerous forces challenging the structural integrity of anatomical structures exposed to discrete impacts. There is today a consensus in the scientific community that vibration values are not relevant to assess the effects on humans of exposure to discrete impacts.vii,viii, ix

SRS, Shock Response Spectrum is a graphical representation of a shock or impact in terms of how a Single Degree of Freedom (SDOF) system, like a mass on a dampened spring, would respond to that input. SRS is developed to be used when a number of mechanical properties of the SDOF are known, and it is also built on assumptions regarding mechanical and elastic properties of a specific defined human body, e.g. the claims regarding resonance frequencies and resonance phenomena.

The relevance of these assumptions has not been shown. To use SRS, the mechanical properties of the tested suspension systems need to be well known as well as the weight of person seated. These properties have just been assigned arbitrary values.

VDV, Vibration dose value is developed to quantify the mean level of energy transmitted as vibration. This unit does not correlate to the actual forces of discrete impacts nor to any physiologic response.x Sed8 has the same limitations.

The concept of creating a unit for describing the severity of exposure to discrete impacts during a time period, whether it is 8 hrs, 8 minutes or 8 seconds, in a single figure, to be used instead of actual acceleration data, has not been justified.

Another flaw rendering all these three units useless for quantifying the forces of discrete impacts, is that they filter out most or all of the energy content above 20Hz.

Comparing the results from seat testing during a three hour transit at normal military operational speed in sea state 3 on subjects seated in fixed seats two subjects seated in state-of-the-art suspension seats, showed highly significant differences in the numbers of impacts and in the amplitudes of the impacts, while in the same data when presented as VDV showed no significant difference. The same study also showed significant difference in the physiological response. The more exposed to group manifested significant physical fatigue, running 386 yards shorter in a shuttle run test, then the less exposed group.

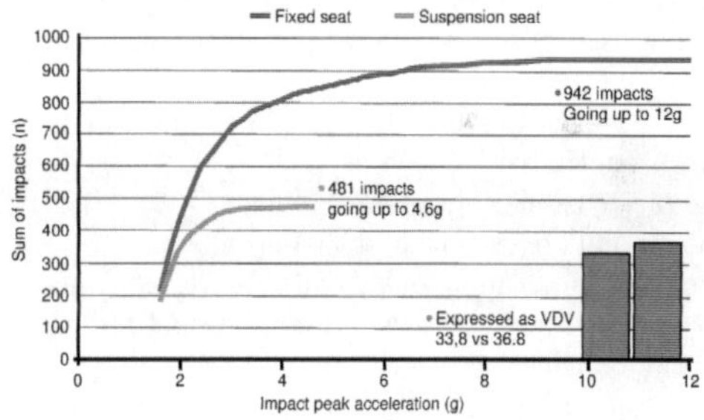

Graph 6. Shows differences in exposure to whole body impact between fixed seats and suspension seats. The subjects on the suspension encountered half as many impacts, 481 vs 942 and only 38% as high maximum peak values. 4.6g vs 12g. Expressed as VDV the difference was insignificant, only 9% or 36.8 vs 33.8 in VDV value.

As the most severe impacts cause the highest risks of injury, it will be necessary to analyse the properties (characteristics) and the energy content of the discrete impacts and the forces they inflict on the body, in order to predict risk for impact-induced injuries.

This observation prompts new evidence based scientific research to do. Such work is ongoing and may bring an understanding of the correlation between measured acceleration data and dangerous forces acting on the body

3.6. "SEATS SHOULD BE PRELOADED WITH DEAD WEIGHTS BEFORE DROPPING THEM INTO THE SANDBOX."
"ACCELEROMETERS SHOULD BE ATTACHED, NOT TO THE DEAD WEIGHT, BUT TO THE SEAT PADDING".
"WEIGHTS SHOULD BE ALLOWED TO MOVE UP TO 50 MM RELATIVE TO THE SEAT PAN AND THE SENSOR".

Justifying claims: *None*.
This positioning of sensors is a methodological flaw.
Acceleration of the isolated part of the seat can differ significantly from the acceleration of the test load representing the human body, and also from the accelerometer between the seat top and the dead weight.
Test data shows that pre-loaded seats, when in free fall, expand their suspensions, accelerating the platform downwards, often faster than free fall.
This acceleration can cause a spatial separation between the seat and the dead weight. The platform impacts with the sand and stops with no dead weight on the seat top. Then the dead weight impacts with the seat cushion.
Test data also shows that the initial impact on the top of the seat suspension often is identical to that of the platform.
When the dead weight then lands on top of the seat cushion, a significant downward acceleration (- g) occurs and gets registered by the sensors mounted on the suspended part of the seat.

With this method, test results will be significantly influenced by a number of interacting factors such as:

- The weight of the suspended part of the seat, relative to:
- the weight of the test-platform including the unsuspended part of the seat.
- the pre-tension of the spring element
- the rebound resistance of the damper
- the slack in fixation and actual relative movement of the dead weight – showing less consistent test results with more slack
- the elastic properties of the dead weight representing the human occupant – soft or hard mass.
- the way in which the sensor is attached to the top of the padding
- the resilience characteristics and the dimensions of the seat cushioning
- the duration of platform deceleration
- the stroke of the seat suspension

All these factors influence the test results in unknown ways and by unknown magnitude, and with unknown relevance for actual performance in real-life exposure with human occupants in the seats.

3.7. PRACTICAL DROP TEST METHOD
The proposed standard does not prescribe a specific method for absorbing the impact of the test platform while impacting. However, two different media have been suggested to simulate the boat hulls impact with the water, sand and aluminium honeycomb blocks. Sand is proposed.
Justifying claims: "*Sand can be easily accessed, reusable and is less expensive than honeycomb blocks which are expensive and not reusable*".
Sand is not useful for this purpose. Only one single set of test was possible to obtain. This shows deviations of up to ± 31% on measured impacts with the same initial setup. This is assumed to be explained by the fact that the energy absorbing properties of sand varies widely depending on type of sand is used, the humidity, how it is packed, by vibration or drop testing, how it is prepared etc.
This wide variation naturally allows for selecting preferred results. No data exist comparing test results from different drop rigs. Already the deviation in the results from one drop rig on one object, proves the method unfit for using as standard. No data allowing assessment of repeatability of drop testing on honeycomb has be presented.

3.8. REPEATABILITY OF TEST RESULTS - REQUIREMENT FOR USE BY ACCREDITED TEST LABS
Justifying claims: No claims were presented justifying any level or lack of repeatability
For a test method to be justified as a standard, it must be possible to use in different locations and achieve basically identical results.
For a test method to be justified as a standard it should also be possible to use, by accredited test labs for certification of function and/or performance.
The requirements for use by accredited test institutes/labs are well defined. The most relevant sources being ISO 17025 General requirements for the competence of testing and calibration laboratories, and ISO17065 Conformity assessment - Requirements for bodies certifying products, processes and services.
All materials, instruments, conditions, occurrences etc that may influence the test results shall be identified and recorded in the test report.
The basis of all qualification testing is repeatable result. Any test, repeated with the same test object, to the same parameters, should give close to identical results.
Therefore; impact testing that uses a perishable material to obtain retardation, is less than ideal, as this entails the need for all test facilities in the world to use the same material, from the one and same manufacturer for all testing.
The use of sand as retardation material is for several reasons not acceptable:

1. Different kinds and different batches of sand contain different particulate size and distribution.
2. Sand with higher contents of fine particles (dust) will absorb more moisture from the surrounding, thus air making sand bed stiffer.
3. Over repeated use, the sand particles will wear down, changing the shape and size of the granules, and the percentage of fine particles will increase.

4. While being repeatedly slammed, sand gets increasingly packed and thereby stiffer. This is well known to roadbuilders and playground designers.

All of the above will make for unrepeatable result and disqualify the proposed method for use by accredited test labs.

3.9. RECOMMENDATION FOR SPECIAL SEAT TYPES - HUMAN INTERACTION

Claim: "*Seats designed to allow bracing positions should not be considered for use and cannot be tested. Bracing efforts can cause physical fatigue*".

Justifying assumption: "*Seats allowing users to adopt bracing positions require bracing, which makes them tired.*"

This is a false claim. No support for this has been found. On the contrary, solid scientific evidence indicates the opposite. Scientific evidence exists showing that some such seats, optimised for synergising with muscular reflex response, without the use of any bracing, can reduce physical fatigue by 31% and impact exposure by 50% in numbers and 62% in peak levels.

The targeted type of seat is the only type, proven by evidence-based science to reduce exposure to, and physical fatigue caused by impact exposure at sea. It is also the only type of seats having a track record of not causing injuries or overboard ejections and not causing overboard ejections.

3.10. ARBITRARY CLASSIFICATION OF BOATS, ANNEX A

A "Test specification by class of craft" was integrated into to the proposed standard.

Justifying claim: Agencies purchasing boats will need help to know if their boats need impact mitigation seats.

Analysis of this classification scheme: This classification scheme defines 7 classes and subclasses, according the authors assumptions of which kind of impact exposure various craft would be able to produce, depending of which kind of agencies are using the boats. These assumptions lack any support in published scientific or experimental data. The classes are vaguely defined which leaves it up the reader to interpret the them. It is not stated how this document can be of any practical use.

It also well known that many suspension seats, which can reduce moderate impacts, can bottom out under more severe impacts, and then actually amplify hull impacts to up to 3 times onto the seat occupants. This means that seats with insufficient shock mitigation capacity can be more dangerous than fixed seats. Therefore it questionable to issue recommendations indicating that certain boats should be OK to be fitted with seats of poorer performance.

3.11. PURPOSE OF STANDARD

Various purposes for use were claimed by the authors of the CD19470. Other purposes where found in advertising.

1. Agencies had expressed a need for guidance in acquiring the best seats that would best protect their personnel from injuries.
2. One agency needed to identify which seats would not be worth testing in real use at sea.
3. One proposal was to determine which seats mitigated and which seats amplified the measured impacts.
4. One producer needed a standard to certify their seats by, protect themselves from liabilities for injuries sustained when using their seats.
5. The same producer claimed that the standard was needed as guidance for other producers in developing more effective "shock mitigating" seats.
6. Recommendations to exclude one certain type of seats from testing and also from use, raises questions about the purpose of the standard.
7. Inclusion of an arbitrary classification scheme without correlation to potential exposure levels, raises questions about the purpose of the standard.

The one purpose for which it was proven useful, and was actually used, was to market a specific brand of seats as the "scientifically proven" safest seat in the world, and to support claims that other specific types of seats should not be considered for use or even tested, and also to strongly advise against any testing of seat performance by comparative testing in real sea trials.

3.12. RELEVANCE OF TEST RESULTS FOR INJURY PROTECTION CAPACITY

A recently published report, comparing suspension seats to rigid seats in respect to transfers of impacts from the deck to seat occupants, shows that rigid seats regularly amplify the deck impacts onto the occupants. This was shown also for suspension seats, but to a lesser extent. The suspension seats could in many cases reduce the impact levels reaching the occupants.

The tested suspension seats were, however, also found to amplify even some modest deck impacts as low as 4-5 g. This was explained by the suspensions bottoming out (reaching end of stroke). The suspension seats were shown to sometimes amplify the deck impacts, even without bottoming out. The report also indicated that the cushioning on both kind of tested seats amplified the impacts, but that this effect could be counteracted to various extents by the suspension mechanisms. The amplifying effect of the cushioning is previously shown in studies on aircraft ejection seats.

Even seats of the same kind as those having shown "excellent performance" in tested according to the proposed lab test method, showed amplification of impacts and could not be verified to reduce human impact exposure at sea to safe levels.

The relevance of the proposed lab test method, for assessing injury protection capacity, was a central issue in the discussions in the ISO Work Group.

The Work Group did in fact establish that the method was not useful to assess the protective capacity of suspension seats.

Based on this conclusion, the convener of the ISO Subcommittee SC4 suggested that the project should be disbanded, and the Work Group dissolved. All nations represented in the Work Group, except for one, voted to disband the project.

4. RESULTS

Analysis of the proposal ISO CD 19470 revealed a number of issues indicating that the proposed method could not serve the claimed purpose to determine seats' capacity to protect users from impact induced injuries.
Several claims and assumptions laid out as the theoretical base for the method have been analysed and shown to be either false or unsubstantiated.
All data from testing done according to the proposed drop testing method is kept secret. No data, even from government funded testing, has been published or made available for scrutiny.
Many of the listed shortcomings were so severe that each one in itself rendered the method useless as test standard.

- Limiting test impact levels to less than half of real exposure levels.
- Avoiding testing for impacts containing horizontal components.
- Failure to produce repeatable results.
- Filtering out the higher frequency parts in measured impacts.
- Converting impact data to vibration data.
- False claims used to issue recommendations for certain types of seats.
- Documented use of results acquired by the claimed " ISO standard" to mislead users.
- Documented use of the "ISO standard" to strongly advise against testing seats in real use in sea trials, claiming this is dangerous and does not give reliable information.
- Lack of correlation between test results and real performance at sea.
- Lack of correlation between test results and protective capacity.
- Disregarding effects body posture and human muscular response to impact.

5. DISCUSSION

This investigation was initiated after the proposed CD 19470 had been passed by a first ballot, to be published as a DIS (Draft International Standard) for a second ballot before final approval as an ISO standard.
Already a year before the first proposal was presented to the ISO committee, it was, however, used in a marketing campaign, leading user agencies to believe that it was in fact an ISO standard, and also that result from this kind of testing were scientifically valid, and that in such a test a specific brand of seats had "unequivocally beat out all competition in every aspect of shock mitigation performance", "overriding all previously published scientific results."
The results of this study show that ISO-standards can be approved even without any support by Evidence based science and that therefore, ISO-standards cannot be assumed, a priori, to qualify as the basis for legislation. Such legislationxi can be pointless or even counterproductivexii and may increase injury risks.
Even the use of such standards for commercial purposes can be counterproductive and may increase injury risks.

In the automotive industry the creation of emission test data, not relevant to real life conditions, has had severe legal implicationsxiii for those responsible severe economic implications for large corporations.
Test results xiv published in October 2019 show that children's car seats when tested to relevant levels of impact are deadly. These seats are marketed with the claims: *All Hauck car seats have received multiple awards and comply with the highest safety standards.* This indicates that standards can vary in relevance and also have different purposes, depending on whether they are designed to protect producers or consumers.

References

---

[i] MASHIEN Study …
[ii] Tom Coe at WG meeting
[iii] KNRM - Rescue Vessel 1816 Hans vd Molen
[iv] Kolla referens 1.4 gånger
[v] MarkV Video + SW
[vi] ISO Standard 2631:1 Mechanical vibration and shock - Evaluation of human exposure to whole-body vibration
[vii] Bovenzi M. Low back pain and exposure to whole-body vibration and mechanical shocks (keynote) 6th International Conference on Whole Body Vibration Injuries WBV2017
[viii] Griffin M Predicting discomfort caused by whole-body vibration and mechanical shock, 6th International Conference on Whole Body Vibration Injuries WBV2017.
[ix] Ullman J. Vibration Exposure Standards are NOT relevant for Impact Exposure 6th International Conference on Whole Body Vibration Injuries WBV2017
[x] Dobbins T. Myers S. High Speed Craft Motion Analysis – Impact Count Index (ICI),
[xi] EU directive 2002/44/EC, on Minimum health and safety requirements regarding the exposure of workers to the risks arising from physical agents (vibration).

[xiii] AUDI emission test adaptation yielded prison sentences for executives,
[xiv] Swedish Consumers' Association, https://www.sverigeskonsumenter.se/in-english/

STRUCTURAL CONSTRUCTION VS WEIGHT SAVING: CHALLENGES IN ALUMINIUM STEPPED HULLS

G. De Mola and **A. Page,** Chartwell Marine, UK

SUMMARY

All vessel users have a responsibility to be mindful of the fuel they are burning, and of harmful emissions resulting from burning such fuel. Compared with traditional hull form technology, a high-speed hull with hull step technology offers the opportunity to reduce propulsive power for the same vessel speed. As a naval architect, the company and/ or individual designing the boat has a responsibility to carefully consider the vessel design strategy. The vessel must fulfil the vessel users' requirements and be designed with hull form and propulsion efficiency in mind from the outset. The vessel should be sustainable, well thought out and be able to be recycled at end of life.

It is therefore important to consider all aspects of the vessel design and build, including the primary materials used for construction. Aluminium alloy could be a suitable candidate to ensure sustainability, however there are a number of considerations which must be taken into account to ensure the use of aluminium is appropriate and effective.

This paper explores the key design process points the naval architect could consider when designing and developing a stepped hull form constructed from marine grade aluminium alloy.

NOMENCLATURE

HSC	High speed craft
Hs	Significant wave height (m)
LOA	Length overall (m)
BOA	Beam Overall (m)
BWL_{HULL}	Beam Waterline Individual Hull (m)
Tc	Draft (m)
$\Delta_{Lightship}$	Displacement Lightship (t)
Δ_{Full}	Displacement Full Load (t)
V_{Light}	Speed Light (kt)
V_{Full}	Speed Full Load (kt)
LCG	Longitudinal centre of gravity
CoG	Centre of gravity
nCG	Dynamic Load Factor
kDC	Design Category Factor
F_{VOL}	Volumetric Froude Number
QPC	Quasi Propulsive Coefficient

1. INTRODUCTION

High speed craft (HSC) are complex. An HSC is subject to ever changing dynamic load forces when passaging offshore. For a "working leisure high speed/light commercial HSC" there can be a variety of operational load cases which must be considered. This is especially evident in high speed sports fishing vessels, typically operated in USA, Australia and New Zealand. In order to travel at high speed, and to be able to achieve this in a variety of operational load cases, typically "high power" is required. High power results in high fuel burn and environmentally harmful by-products (gases) emitted. As we look to reduce the environmental impact, can the operation of this vessel type still be justified? Can this operation be made sustainable, and what can we learn from it to apply to military and commercial applications?

This paper presents a high-speed 13 m light commercial sports fishing catamaran. This HSC uses a stepped hull form, with a view to allowing the operator to travel at high speed, but at a lower fuel burn rate than a traditional unstepped hull form. The vessel is designed to be manufactured in aluminium. It is to be efficient, cost effective and capable of operating offshore. It is to be able to be recycled at end of vessel life.

2. DESCRIPTION OF VESSEL

The vessel is a 13 m (42 ft) high speed light commercial fishing vessel, for sports fishing purposes, designed to be operated off the east coast of the USA, between Newport and Martha's Vineyard, Rhode Island. This HSC will be capable of operating at speeds in excess of 45 knots, in sea states of approximately 1 m Hs. The vessel is designed to support a variety of payloads, with onboard tankage provided for carrying fuel, fresh water, black water, bait tanks and catch tanks. The vessel is arranged with a wheelhouse and flybridge. A total of eight (8) persons may be carried onboard.

Based on the defined operational profile and variety of payloads, it was determined that a catamaran hull form would be appropriate. The catamaran provides a transversely stiff hull form, capable of supporting differing payloads without experiencing high angles of heel. Furthermore, for the purpose of sports fishing, a catamaran provides a wide-open deck with space for personnel and fitment of fishing equipment.

There are a number of possible options when designing the hull forms for a high-speed catamaran. It is recognised that asymmetric hull forms, characterised by wall sided inboard, and medium/large angle deadrise outboard, offer a low resistance high speed hull form. However, an asymmetric hull form design carries certain disadvantages. It does not provide as much volume as a symmetric hull form for load carrying. Furthermore, dynamically, the vessel will experience a greater angle of heel when undergoing a high-speed turn than a symmetric hull form. Therefore, given that this vessel is to operate with differing payloads and to perform offshore in seas of 1 m Hs, a symmetric hull form was selected as the most appropriate option. The main specifications for the vessel are:

- LOA: 13 m approx.
- BOA: 4 m
- BWL_{HULL}: 1.61 m max.
- Tc: 0.74 m approx.
- ΔLightship: 10.2 t approx.
- ΔFull: 15.52 t approx.
- VLight: 55 kts
- VFull: 45 kts
- Construction Material: Marine Grade Aluminium.

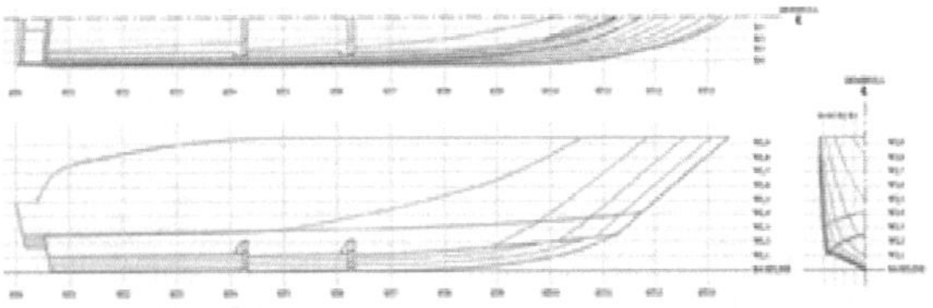

Figure 1: Lines plan for 13 m light commercial sports fishing catamaran.

When designing a high speed vessel, in order to maximise hull form efficiency, the following hull form appendages may also be taken into consideration:

- **No steps**, traditional symmetric hull form, clean bottom
- **One (1) step** – a single step placed at or close to the anticipated operational LCG
- **Two (2) steps** – 2 steps placed either side of the anticipated LCG
- **Three (3) steps** – 3 steps equally spaced in the proximity of the LCG
- **Transom Interceptor** – a fixed or adjustable vertical blade or panel placed at the transom, perpendicular to the hull bottom and water flow
- **Trim tab(s)** – an adjustable plate (or plates), placed parallel to the hull bottom at the transom. The trim tab is adjusted such that the angle of attack generates transom lift as the water flow passes under the hull bottom beyond the transom
- **Hydrofoil** – a foil profile, mounted transversely under the hull bottom, midship and/or aft, positioned appropriately about the LCG to generate lift, and, if adjustable, to control/ dampen vessel dynamics.

For the vessel design presented in this paper, hull steps were chosen as the primary hull appendage, integrated with a transom interceptor, to provide dynamic adjustment to suit the varying payloads.

3. HULL DESIGN/ GEOMETRY

For a high-speed craft, hull bottom deadrise angle and chine design is critical to performance, both in terms of speed and seakeeping. When designing a stepped hull, deadrise design can be considered even more critical, where we are trying to interfere with flow and fluid density and how they interact with the hull surface. In the case of the 13 m catamaran deadrise influences the following:

- Performance
- Seakeeping capability
- Boat handling and operation
- Aluminium hull form plate formation.

Typically, when defining hull bottom deadrise, a compromise has to be made between resistance and seakeeping. The main benefits of high deadrise hull bottom are good directional stability at high speed and lower vertical motion amplitude. However, it also entails a series of drawbacks:

- Higher resistance at slow/pre-planning speed due to increased transom immersion
- Higher resistance at high speed due to reduced hull surface area, resulting in lower lift force at transom.
- Low speed manoeuvrability can be reduced
- Reduced transverse stability at high speed (although this is more applicable to monohulls).

As the primary function of this vessel is to operate at high speed efficiently and safely in up to 1 m Hs offshore, a fairly shallow, low deadrise angle was chosen. The decision-making process on deadrise angle and how it interfaces with the hull step design are further detailed in section 4.4 of this paper.

4. HULL STEP DESIGN

A stepped hull is characterised by one or more "steps" which are fitted transversely from the hull keel to the chine(s). Dynamically, when water flow passes over the step, it finds a vacuum zone (pocket of air) which results in a reduction in wetted surface and therefore a reduction in hull form frictional resistance. If designed correctly, the result of this reduction in frictional resistance is an increase in hull speed for the same given installed propulsive power.

4.1 TYPICAL STEPPED HULL CHARACTERISTICS

A hull step is a transverse "notch"/ recess in the hull bottom (Figure 2). It intersects with the hull topside, in such a way that it results in a ventilation opening just above the waterline when the vessel is operating in planing mode. As the vessel moves through the water, a pocket of low pressure is created just aft of the hull step. The low pressure creates suction that draws air in from the ventilation opening. Since speed is a function of lift, as the speed increases, so does the lift, thus causing more air to pass through the ventilation opening, which then separates the hull bottom aft of the step from the water. Since density is a function of drag, the drag is reduced, as air is in contact with the hull bottom aft rather than the water. As a result, the vessel is able to passage using less installed power than an equivalent unstepped hull form.

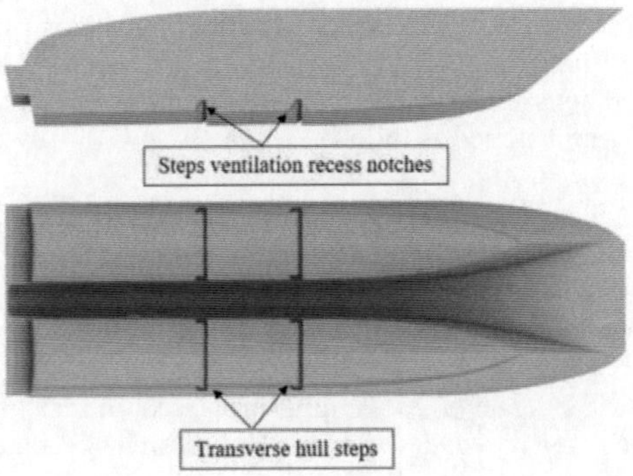

Figure 2: Example Multihull with hull steps

A multistep hull form can be determined appropriate, depending on the vessel's length to beam ratio, and the required vessel speed. For a vessel with a "high aspect ratio" i.e. narrow beam for its length, it is typical for two steps to be fitted. In fact, two steps were chosen for the 13 m high speed light commercial sports fishing catamaran, to serve two primary purposes: increased reduction in frictional resistance and a shift forward (from the transom) of the dynamic pivot point when travelling at high speed.

4.3 STEP(S)/ LCG POSITIONING

As stated in section 4.2 of this paper, the incorporation of hull steps leads to a shift in the longitudinal pivot point from the transom to the hull step. If the vessel is designed correctly, both in terms of step design and LCG location, the incorporation of hull step(s) can lead to a reduction in vessel trim/ pitch (Figure 3).

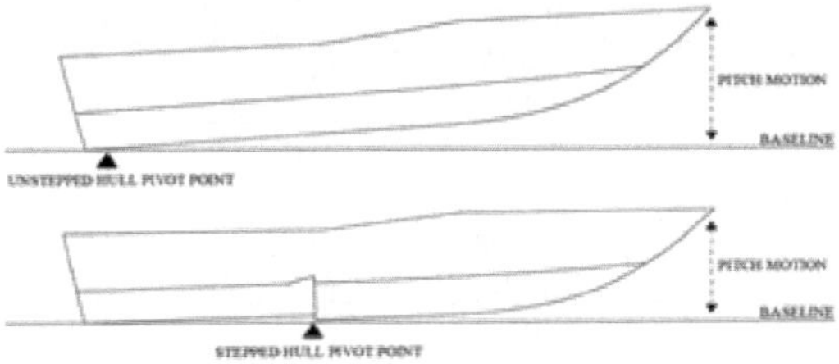

Figure 3: Example of stepped and unstepped hull forms, comparing the typical hull form pivot point and pitch motion.

The LCG should be positioned about the hull step(s) such that the steps do not cause an undesirable dynamic moment which could make the vessel unstable. For a single step hull form, it is typical to position the LCG aft of the hull step. If positioned close to the hull step, the pitch gyradius is low. If positioned "far away" the pitch gyradius is high which may result in a higher pitch moment, which could cause the vessel to porpoise. For a two (2) step vessel, it is typical to position the LCG just forward of the second step, aft of the forward step (Figure 4).

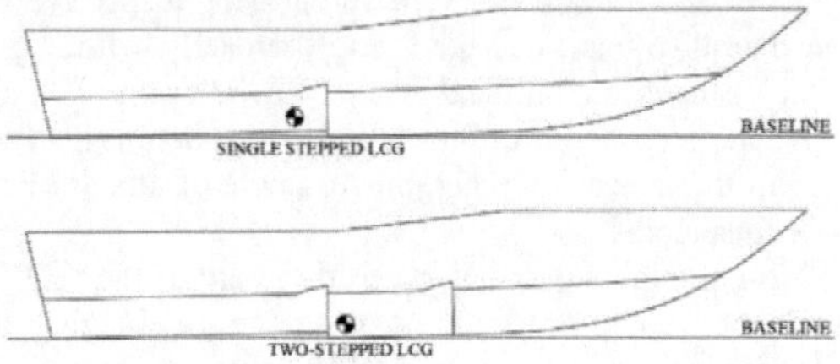

Figure 4: Example of single vs two-stepped hullform LCG position

As well as positioning the LCG about the hull steps, it is also critical to correctly position each step at a correct distance from the other. If the steps are positioned too closely together, the flow which has passed the first step will not have time to reattach to the hull bottom before it reaches the second step. Were this to occur, the water flow passing over the second step would be turbulent (a mixture of water and air), which means the second step would not generate as much lift as it would if the flow were laminar and single density. The second step would therefore be redundant, or worse, lead to the hull form being less efficient than if it were un-stepped.

4.4 HULL STEP/ HULL DEADRISE INTERFACE DECISION- MAKING

Peters [1], finds that the deadrise angle between the transom and the forward step should not vary by more than 2 degrees, to ensure good flow reattachment between the first and second steps. The sports fishing catamaran hull form therefore has a slight "warp" in hull form aft, limited to a maximum of 2 degrees. Forward of midships, the vessel has a relatively high deadrise. It is anticipated that this will provide a lower buoyant volume forward with respect to bow amplitude (pitch) and also a lower bow profile driving towards the prevailing wave in order to minimise wave resistance. It is anticipated that this arrangement should result in advanced hull efficiency in the planing length of the craft, and reasonable dynamic seakeeping capability, in pitch, heave and wave dissipation.

In order to qualify the decision-making as expressed above, Chartwell Marine commissioned the build of two (2) 1:12 scale hull models. Each hull has been tested independently by Seaspeed Marine Consulting at QinetiQ Haslar towing tank facility, UK, to establish the hull form resistance, dynamic trim and sinkage, and its dynamic high-speed stability. A summary of results from this testing can be found in chapter 7 of this paper.

4.5 STEP SECTION (PROFILE)

Following the findings of Peters [1] and Garland [2], the authors determined that step height should be between 31.8 mm and 65.5 mm and approximately 4% of chine beam, respectively. For the 13 m catamaran, a step height of 36 mm and 50 mm (forward and aft step respectively) have therefore been designed, approximately within the optimal ranges as outlined above. In addition to step height, the step angle of attack should be determined to conform the chosen step height. An angle of attack of 1° has been selected.

Choosing the ideal parameters requires careful thought to find the best compromise. In our case, the step height and angle of attack chosen are considered sufficient to provide the required vessel performance. At the same time, their design is considered achievable as regards the feasibility of manufacturing in aluminium alloy.

4.6 CHINE DESIGN & HULL VENTILATION

In order to ensure good ventilation to the hull area directly aft of the hull steps, the size and shape of the aperture recess/ notch in the hull bottom/ topside (chine) interface is important. If a chine notch/recess is too small or incorrectly positioned the airflow will be ineffective. Furthermore, if "blocked" by water, a vacuum may be created in the step notch/recess at the vessel chine/side which can result in an undue heeling moment which can warrant the vessel unstable. It is important that the notch is arranged so that continuous air flow is maintained, both in flat water and in waves.

For the 13 m catamaran a reasonably large notch has been identified as appropriate. This is because the vessel is to operate in a number of different load cases and a number of different sea states. Moreover, since the vessel is to be constructed in marine grade aluminium alloy, it is important that there be adequate space for plate formation and welding, to ensure hull water tightness and structural integrity.

5.0 VESSEL STRUCTURAL DESIGN

The layout of the structural design arrangement, and respective structural weight, are critical aspects of working leisure HSC/light commercial HSC craft subject to frequent variations in their daily operational payloads. Therefore, the authors have taken into consideration the following main points for the structural design loop of the design spiral process:

- Selection of suitable vessel build material
- Identification of the most apt structural design regulations for the project
- Identification of the vessel's typical loading conditions & operational payloads
- Comparison analysis of different design pressures vs operational load case scenarios
- Optimization of structural arrangement: minimum scantling requirements vs overall structural weight
- Vessel build philosophy, looking at the challenges posed by the design geometry of stepped hull form(s), the shipyard technical capabilities/expertise, and the resource budget allocated to the vessel build itself.

5.1 OPERATIONAL CONSIDERATIONS

To correctly identify the loading condition scenarios and their respective overall displacements, the main considerations to be made for the vessel's operational profile were outlined as follows:

- It will typically perform charter and leisure fishing operations
- It will be able to carry out fishing operations in sea state conditions up to 1 m Hs, in sea regions up to offshore
- It will operate at 35 kts cruising speed and 55 kts maximum speed, based on market research of similar competitor vessels and client requirements.

The overall displacements for each individual loading condition of this design (based on the above points and taking into consideration industry common practices and specific client requirements with regard to operational payloads) are summarised in Table 1.

Table 1: 13 m Light Commercial Sports Fishing Catamaran Operational load cases.

LOADING CONDITION	DESCRIPTION	DISPLACEMENT MASS (t)
1	Full Load Departure	15.516
2	On Passage / Fishing	13.705
3	Arrival with 90% Fuel & Fish Product Caught	15.142
4	Arrival with 10% Fuel & No Fish Product Caught	11.941
5	Arrival with 10% Fuel & Fish Product Caught	12.058
6	Arrival with 10 % Fuel & Extra Fish Product Caught	12.207
7	Charter Fishing Trip Arrival with 60% Fuel	13.926
8	Sea trials	11.368

5.2 VESSEL BUILD MATERIAL

In order to select the most suitable vessel build material, it is important to consider its influence on both the structural design and its respective weight.

The selection of marine grade aluminium alloy for this project, as opposed to composite, might therefore seem difficult to justify if comparing kg for kg aluminium vs composite. Aluminium however offers the following advantages:

- Repairability and ease of maintenance throughout the operational life cycle of the vessel, a key factor to consider when designing commercial/charter type craft with continuous upkeep costs
- Structural strength and rigidity
- Competitive build production costs
- Ease of customization; adding a new feature or section to an aluminium vessel is fairly straightforward, whilst composite requires the fabrication of a moulding tool to achieve the same
- Endless material recyclability, another major advantage of aluminium over composite, with the increase of market demand for sustainable and environmentally friendly designs, aluminium lends itself particularly well to this objective.

5.3 APPLICABLE DESIGN CRITERIA ISO 12215-5

For the purposes of this paper ISO-12215 (2008) [3] regulations were utilised to design the vessel scantlings. Although these regulations do not yet provide a comprehensive set of design guidelines for multihulls, it was determined appropriate for use, assuming half the maximum fully laden vessel displacement value to input in the design pressure calculations, as required in the ISO 12215 regulations. The other required hull dimension data were treated as if the individual hull of the catamaran design were a monohull (see below). This provided sensible minimum scantlings in line with case studies researched for this project. The specifications for this vessel are:

- Displacement - m_{LDC}: 7760 kg
- Waterline Length – LWL: 11.044 m
- Chine Beam – B_c: 1.44 m
- Deadrise Angle – ß: 20.8 deg.
- Maximum Design Speed – V: 55 kts.

5.4 DESIGN PRESSURES

A comparative analysis of the maximum design speed input vs the design pressures for minimum scantling requirements was performed, with respect to the selected operational vessel design categories as defined by the ISO-12215 (2008) regulations. This analysis provided the authors with the opportunity to identify a "worst case scenario" for the structural arrangement design, whilst still pursuing the objective of an optimized scantling design.

Dynamic Load Factor (nCG) is the ruling factor for determining base design pressure values for planing craft, as it is directly proportional to the maximum speed input value. Table 2 summarises the results for 3 operational profile scenarios.

Table 2: Vessel particulars data input for ISO-12215 (2008) criteria.

ITEM DESCRIPTION	Condition 1 - 10 kts CAT B	Condition 2 - 30 kts CAT C	Condition 3 - 55 kts CAT D
Dynamic Load Factor - nCG	0.212	1.912	5.999
Design Category Factor - kDC	0.8	0.6	0.4
Base Bottom Displacement Pressure - $P_{BMD\ BASE}$ (kN/m2)	66.117	66.117	66.117
Base Bottom Planing Pressure - $P_{BMP\ BASE}$ (kN/m2)	58.068	121.070	233.917
Base Deck Pressure - $P_{DM\ BASE}$ (kN/m2)	18.465	18.465	18.465

The results highlight the fact that the hull bottom region of the vessel is the area subject to the largest base design pressure values, with the Condition 3 scenario yielding the highest; this condition provides therefore the worst-case scenario for the structural calculations.

5.6 BUILD METHODOLOGY

To design a strong and lightweight structural arrangement, an understanding of the relationships between the individual structural items and the overall scantlings' matrix is required. The structural arrangement must also be logical in its layout in order to optimise it without compromising its integrity. To achieve this, the authors focused their analysis on the scantling matrix that forms the basis of the structural arrangement of a vessel (Figure 5).

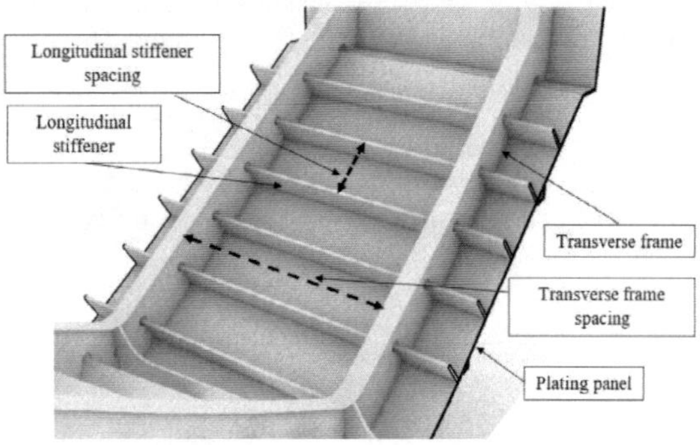

Figure 5: Scantling matrix

Using both the results of this analysis and a knowledge of fabrication and construction assembly of aluminium vessels, the authors were able to make the following observations on which to base the build philosophy/strategy:

- **Plating thickness** - The main influencing factor for plating thickness is the longitudinal stiffener spacing. A thicker plating thickness (based upon the authors' analysis of the calculations as per the adopted structural design criteria), will allow for the secondary structural components of the scantling matrix to be spaced further apart; this solution normally requires less design and build "effort"; it will be referred to as METHOD 1. Instead, a thinner plating thickness will require more time to achieve a suitable compromise between strength, weight of the internal structure (excluding plating), design effort and ease of build; this design approach has been named METHOD 2.
- **Stiffener spacing** - The main considerations for longitudinal stiffener spacing are the impact it has on minimum required plating thickness; the influence it has on transverse frame spacing and finally, the impact it has on vessel construction (welding accessibility).
- **Frame spacing** – Although it may seem that the easiest solution would be to design a secondary structural scantling matrix with closely spaced transverse frames, it is not necessarily the most efficient solution with regards to cost (more frames will require more material and CNC cutting, with additional labour hours to assemble), nor is it the most effective in achieving a fair and smooth outer hull plating shell, due to the welding deflection which can be very appreciable in aluminium vessels. This is an important consideration to be made particularly when designing leisure craft, where aesthetics plays a central role. A visible "ripple" effect due to close transverse frame spacing could require additional fairing compound to be applied to the plating; this in turn will result in increasing the overall lightship weight.
- **Framing vs longitudinal stiffeners web depths** – Larger longitudinal stiffener web depths will result in overall larger minimum transverse framing web depths, to accommodate the recess notches in the transverse frames for the longitudinal stiffeners to be continuous. It is common practice to design transverse frames to have a minimum web depth which is double the height of the longitudinal stiffener.
- **Shipyard capabilities & supply chain challenges** – A shipyard's workforce skillset and the client's "purchasing budget" for the project are other factors to consider. The cost of purchasing thinner shell plating (therefore less material) or standard stock extrusions for the secondary stiffeners are likely to result in a lighter structural weight, compared to a design with heavier shell plating or many CNC cut parts. However, the former is likely to require a higher skilled workforce plus additional labour hours to complete the intricate fabrication assembly of an aluminium stepped hull form. Finally, it is useful to determine early on, if possible, the likely supplier for the vessel's build material and identify any "potential challenges" there might be including: range of plate thickness available, CNC machining capability and availability of extrusions. A lack of any of the above might force both naval architect and shipyard to potentially re-think their build philosophy strategy.

5.7 STRUCTURAL DESIGN OPTIMISATION

As can be seen in Table 2, the structural items forming the hull bottom region scantling matrix are required to withstand the highest design pressures. The hull bottom region structural items therefore make up a significant portion of the overall structural arrangement weight. As a result, the structural design optimisation process is linked to the design of a suitably strong lightweight hull bottom region structural arrangement.

To achieve this, the overall bottom region structural weight was calculated for the different build methodology options outlined in section 5.6. These are compared in Table 3.

Table 3: Vessel structural weight calculations for build methodology options 1 and 2.

SCANTLINGS	METHOD 1	METHOD 2
Hull Bottom Region Plating Thickness (mm)	8	6.35
Hull Bottom Region Plating Area (m2)	51.05	51.05
Hull Bottom Plating Mass (kg)	1102.68	875.25
Transverse Frame Spacing (mm)	1000	500

No. Transverse Frames (#)	13	23
Transverse Frame details	220 x 6 mm WEB 40 x 4 mm RIDER BAR	130 x 4.76 mm WEB 50.8 x 4.76 mm RIDER BAR
Transverse Frames & Rider Bars Mass (kg)	462.79	469.60
Bottom Region Longitudinal Stiffener Spacing (mm)	300	200
No. Longitudinal Stiffeners per hull (#)	4	6
Tot. Overall Length Longitudinal Stiffeners (m)	86.22	126.13
Longitudinal Stiffener details	110 x 10 mm FB	65 x 6.35 mm FB
Longitudinal Stiffeners Mass (kg)	256.07	140.56
OVERALL STRUCTURAL MASS (kg)	1821.54	1485.42

The structural arrangement weight calculated with Method 2 is approx. 20% lighter than with Method 1, highlighting the importance of the design refinement process to achieve the desired structural optimization. The reduction of the required structural material also confirms that an optimised, and consequently lighter, structural arrangement achieves the purpose of a more sustainable HSC design in aluminium alloy.

6.0 WEIGHT ESTIMATE ANALYSIS

A detailed weight estimation analysis is instrumental in defining the operational loading conditions of a vessel and their respective LCG positions, a critical factor for stepped hull forms and high speed commercial craft with many variable payloads.

To best understand the weight and weight distribution, a process of weight estimation control must be implemented. This can be achieved by comparing displacement values of competitor vessels (parametric study), keeping an accurate updated 2D/3D model of the design and most importantly, a good understanding of the influence that the individual heaviest/variable weight components (tankage, structure, cargo payloads, propulsion system, etc.) have on the overall CoG positions across the range of different loading conditions.

6.1 WEIGHT OPTIMISATION

Weight optimisation is not necessarily limited to a weight reduction exercise on the structural design arrangement only. The weight optimisation process is also comprised of the identification of the heaviest items onboard, their respective "safety margin" (a percentage of the total weight of each item) and the verification that the initial weight value (received or researched) is appropriate.

Thorough research and careful thought need to be given to all aspects of the weight estimation. For this project, in addition to the structural optimisation process results discussed above, further research was performed to obtain information regarding the following major items:

- **Fuel tankage** – due to the operational range and speed requirements of the client, fuel consumption and consequential tankage size and mass were critical. Tank testing, sea trial data, parametric study of competitor vessels and engine manufacturer performance data were all tools used to validate the final tankage size.
- **Fish cargo storage/tankage** – this is a key factor to consider for high speed sports fishing vessels, even more so if a stepped hull form is to be used. Because of the operational requirements, large fish storage tankage volumes are to be installed both above and below the main deck across the length of the hull. Therefore it was important to have direct input from the client to define their exact installation location and their intended use in terms of fill percentage of water and/or fish cargo, thus guaranteeing the least LCG position variation across any of the operational payload scenarios outlined in section 5.1.
- **Noise & vibration dampening insulation** – as with the fuel tankage, the weight of the noise & vibration insulation for the hull and superstructure was another important matter to consider. The original specification supplied by the manufacturer accounted for the insulation to be approx. 12% of the overall lightship weight; however, after careful research, an accurate 3D model replica of the vessel areas to be treated and in-depth discussion with both client and manufacturer this was reduced to 7%.
- **Paint application** – due to the aesthetic requirements of the client, the vessel requires a very high-quality finish involving a high spec paint application matrix, including allowance for significant use of fairing compound in the weight estimation. The authors had dialogue with industry experts, and as such were confidently able to input a weight margin budget against the paint application for this vessel. Further, following extensive correspondence with paint experts and marine shot blasting companies, the authors were able to determine the 100% recyclability value of the entire vessel structure, both shell plating and secondary structure at vessel end of life.

Weight optimisation also includes identifying what actual items the client requires in order to carry out the outlined operational demands of his/her vessel. To this purpose, the naval architect must engage an open-minded conversation with the client and explain the need for potentially discarding certain items or "add-ons" that do not necessarily add engineering or operational value. For example the installation of a diesel marine generator and the respective diesel fuel tankage for its running onboard, might seem difficult to justify when any electrical hotel loads can be sufficiently met with the use of alternators running off the outboard engines, a technological solution that is rapidly gaining in popularity across the outboard engine and high speed sports fishing vessel markets. Moreover, a decision to build certain structural components (considered as being above the main deck and not integral to the vessel itself) in lightweight composite rather than aluminium, is another potential solution towards a lighter or at least a more "controlled" design weight budget-wise.

7.0 MODEL TESTING RESULTS

As mentioned in section 4.5, to validate the design decisions made with regards to hull form geometry, the authors commissioned model testing of two (2) sets of hulls, in addition to software resistance prediction calculations.
The main characteristics of the hull forms produced for the tank testing are:

- **Model 1 -** (two steps) - 31° deadrise angle at transom, 13.42 t displacement, LCG 4.73 m forward of zero point, 1.34 m max BWL individual hull.
- **Model 2** - (two steps) - 22° deadrise angle at transom, 15.142 t displacement, LCG 4.894 m forward of zero-point, 1.61 m max BWL individual hull.

It can be seen that "Model 2" has a greater hull volume than "Model 1". This is to give the client more flexibility in carriage of payloads.

In addition to the above, the authors also compared two further hull forms, with the same characteristics as Models 1 and 2, but without steps. These two (2) unstepped variants were compared using the conventional Savitsky prediction method for planing hulls; their main characteristics are:

- **Model 1 - No Steps** - 31° deadrise angle at transom, 13.42 t displacement, LCG 4.73 m forward of zero point, 1.34 m max BWL individual hull.
- **Model 2 - No Steps** - 22° deadrise angle at transom, 15.142 t displacement, LCG 4.894 m forward of zero point, 1.61 m max BWL individual hull.

The tank test and the Savitsky prediction calculations for the different hull forms resistance results are shown in Figure 6.

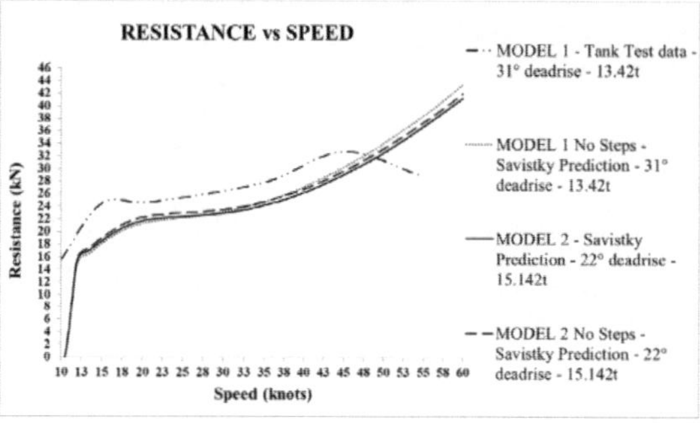

Figure 6: Model 1 vs Model 2 hull form resistance comparison, stepped vs unstepped.

7.1 HULL STEP BENEFIT (POWER REDUCTION)

Unfortunately due to unforeseen global circumstances, the authors were unable to perform the tank testing for Model 2 at the time of writing this paper, however from the research carried out thus far and the resulting resistance curves summarised in Figure 5, the following main observations were made:

- The "Model 1" stepped hull form at 35 kts and 45 kts yields respectively an 11% and 8% increased total resistance compared to the "Model 1 - No Steps" design. However, at speeds greater than 45 kts (F_{VOL} 5 approx. upwards), the stepped hull form design offers a resistance reduction of up to 33% at 55 kts (F_{VOL} 6.2 approx.). Using a 55% QPC value, the result is a 25% power reduction for the stepped hull form design to achieve the maximum speed (55 kts) compared to the unstepped hull form.
- From the comparison of the Savitsky predictions for "Model 2" vs "Model 2 – No Steps", the stepped hullform design offers a 2 to 3% resistance reduction compared to the unstepped design. This trend commences at speeds greater than 35 kts (F_{VOL} 3.8 approx.).
- From a comparison of the Savitsky predictions for "Model 1 – No Steps" vs " Model 2 – No Steps", it can be observed that the "Model 2 – No Steps" with a lower deadrise angle, yields a resistance reduction of up to 3% at 55 kts.

It is planned that model 2 will be tested at the QinetiQ Haslar Tow Tank in July 2020.

7.2 IMPACT OF HULL FORM GEOMETRY, DEADRISE AND LCG

Based on the considerations outlined in the previous section, the following main design factors have been reviewed:

- **Hull steps**: the proposed hull form geometry, a two (2) stepped catamaran design, offers superior resistance efficiency at speeds above 45 kts. Ultimately, at the outlined maximum speed of 55 kts for this project, a stepped hull form will potentially require 25% less power to achieve the same speed (based on the "Model 1" tank test data vs the Savitsky prediction results).
- **Deadrise angle**: Although the resistance increase at pre-planing speeds is mainly attributable to the increased displacement of "Model 2 – No Steps", it is likely that this is partially offset by the change in deadrise angle aft compared to "Model 1 – No Steps" as a result of reduced transom immersion. In addition, "Model 2 – No Steps" offers a resistance reduction at speeds greater than 35 kts (F_{VOL} 3.8 approx.), which is attributed to the greater surface area and consequential lift generated as a direct result of having a lower deadrise angle.
- **LCG**: the forward shift of the LCG position of "Model 2" compared to "Model 1" yielded no appreciable resistance increase or reduction (stepped or unstepped).

8.0 CONCLUSION

8.1 PERFORMANCE CONCLUSION

It can be seen that although weight optimisation is clearly important, hull form geometry and design, in this example, appears to be more significant. As can be seen in section 7.0, although "Model 1" operates with a lower displacement than "Model 2", the revised hull geometry, with reduced hull deadrise aft and increased hull beam results in a reduction in transom immersion, beneficial at slow speed, and an increase in aft lifting surface, beneficial at high speed. This is very significant and an important consideration for naval architects when considering hull form design for multi payload carrying vessels.

For this vessel, assuming the maximum displacement condition, as defined by the operational loadcase by the client, the aluminium structure accounts for 20% of the total displacement. Although, it could be possible, potentially to reduce the weight of the hull form structure using composite technology, it is considered that in this instance, providing the client with a detailed operation manual, whereby they are advised to not carry payloads unnecessarily, will have a much larger influence on hull form resistance and therefore required power and resultant fuel burn.

If we are to make more sustainable vessels for the future, then it is the responsibility of all vessel stakeholders to be mindful of the way they are using the vessel.

8.2 SUSTAINABILITY CONCLUSION

From the results presented in this paper, the authors are confident in their findings, and highlight the following:

- Although it was found that at pre-planing and slower planing speeds a stepped hullform has a marginally increased resistance compared to an un-stepped hull, this type of design lends itself particularly well to light commercial high speed vessels with a continuous high- speed operational profile. In fact, for the purposes of this project, the designed stepped hullform provides superior performance at speeds above 45 kts, with a 25% power reduction to achieve the set maximum speed (55 kts).
- The selection of marine grade aluminium alloy has been found to offer several advantages over composite as the chosen build material. Aluminium not only offers benefits for ease of build and repair, it is considered a more sustainable material choice than composite. In fact, the entire aluminium structure can be recycled including the shell plating, thanks to the use of modern paint products coupled with environmentally sustainable paint application removal techniques.

Taking 8.1 and 8.2 into account the authors of this paper determine that it is possible to design an efficient and sustainable multi-payload high speed vessel using recyclable marine grade aluminium. It is appropriate for sports/leisure, workboat and military applications.

9.0 ACKNOWLEDGEMENTS

The authors would like to thank Seaspeed Marine Consulting, Soundown Corp. and International Yacht Paint for the technical support provided. In addition, the authors thank Francesco Chiviló, head of design at Chartwell Marine, for design support provided throughout the course of this project.

10.0 REFERENCES

1. PETERS, M., 'Peters On (Fast) Powerboats. Part 2', *Professional BoatBuilder,* 2010, pp. 56-71 Oct/Nov issue.
2. GARLAND, W. R., 'Stepped planing hull investigation', *United States Naval Academy*. 2010.
3. INTERNATIONAL ORGANIZATION FOR STANDARDIZATION, 'Small craft — Hull construction and scantlings — Part 5: Design pressures for monohulls, design stresses, scantlings determination', *ISO-12215*, (2008).

9. AUTHORS BIOGRAPHY

Guido De Mola holds the current position of Marine Design Engineer at Chartwell Marine Ltd. He is responsible for small craft design. His previous experience includes project management and sea trials duties within the commercial maritime vessels refit field.

Andy Page holds the current position of Managing Director and Naval Architect at Chartwell Marine. He is responsible for the management and oversight of design and engineering at Chartwell Marine. His previous experience includes being the designer of multiple high-speed craft, including crew transfer vessels, fast ferries, and high-speed rescue boats.

SUPER SWIFT SERIES – THE SOLUTION TO COASTAL DEFENSE

W. H. Kenneth Lee, ST Engineering Marine Ltd, Singapore

SUMMARY

Unpredictable maritime threats have influenced the operational requirements for coastal surveillance. The ability to respond quickly to achieve higher sustained cruising and sprint speeds, typically in excess of 30 knots cruise and beyond 60 knots sprint is evident for any interceptor or fast patrol boat. In response to these requirements, the Super Swift series fast patrol boat is one offered by the Marine sector of ST Engineering in meeting these demands.

At ST Engineering, the Marine sector has proven design track records in combining the state-of-the-art hydrodynamic design methodologies to develop a wide range of planing hull forms meant for ultra-high-speed applications, to achieve in excess of 60 knots, which can be extremely challenging for any 50-tonne craft.

The main objective of this paper is to present a novel approach on the hydrodynamic aspects of various in-house developed planning hull forms. In addition, the paper will also demonstrate optimisation technique adopted to further enhance human ergonomics, crew comfort and achieve multi-role capabilities.

NOMENCLATURE

B	Beam of hull (m)
Δ	Displacement (t)
g	Acceleration of gravity (m/s^2)
L_{CG}	Centre of gravity (m)
L_{OA}	Length overall (m)
L_{PP}	Length between perpendiculars (m)
V	Speed of craft (m/s)
T_f	Draft Forward (m)
T_a	Draft Aft (m)
∇	Displacement volume (m^3)

1. INTRODUCTION

Maritime threats, which include piracy and maritime terrorism, affect the stability of national and international economies. As shown in Figure 1, the number of piracy and armed robbery against ships in Asia has been rapidly increasing. 82 incidents of piracy and armed robbery against ships were reported in Asia in 2019; an 8% increase in the total number of incidents and a 15% increase in the actual incidents compared to 2018[1].

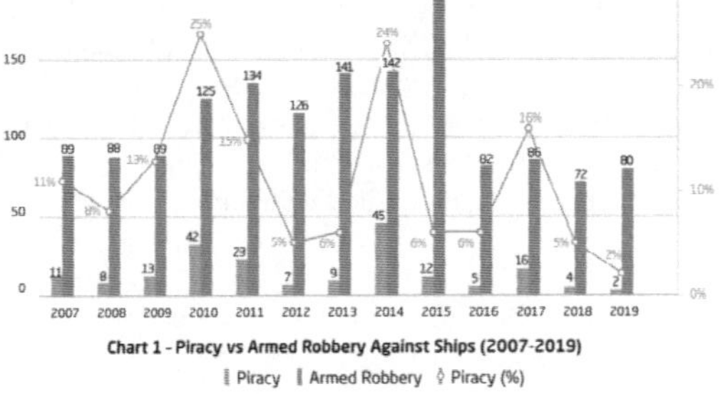

Fig. 1. Number of piracy and armed robbery against ships in Asia from 2007 to 2019 [1]

In an environment where threats are sudden and unpredictable, it is paramount for frontline coastal defence forces to be well equipped with new capabilities for swift response. ST Engineering understands the need for addressing these and aims to provide a solution to such.

2. PROBLEM STATEMENT

ST Engineering embarked on this journey by identifying the various shortfalls of conventional high-speed craft designs based on past experiences derived from its subsidiaries in Singapore (ST Engineering Marine Ltd) and US (VT Halter Marine, Inc.).

High-speed vessels are nothing new to ST Engineering, with remarkable and notable design and build track records. In recent years, the focus has turned to various novel craft designs as naval architect continuously seek to design higher speeds craft by identifying means to reduce the wave-making resistance. Various hull forms that were considered for the development of the Super Swift series include the stepped planing hull, advance deep-vee planing hull, enhanced hard chine planing hull and air cavity hull.

The objective for this development is to cater for different types of operations, demands and providing alternatives to meet stringent operating requirements.

2.1 LOW DISPLACEMENT TO LENGTH RATIO

It is typical for a conventional high-speed craft to have low displacement/length ratios to achieve high-speeds. As a consequence, their motions and behaviour can be affected noticeably in severe sea states. [2]

As such, two key design aspects for high-speed crafts are compromised; seakeeping performance and the ability to equip themselves sufficiently to counter marine threats.

2.1.1 SEAKEEPING PERFORMANCE

Low displacement/length ratio makes conventional high-speed craft more vulnerable to behavioural aspects such as broaching, surfing and bow diving phenomenon [4]. It could even lead to devastating impact such as injury and/or in an unfortunate event, loss of life.

Such limitations and operating speed requirements have led to high vertical acceleration forces of up to 10g, which was reported on board the MK V SOC, causing severe injury due to high exposure to shock and vibration, causing discomfort, injury and performance degradation to the crew [3].

In planing mode, majority of a craft's weight is supported by hydrodynamic pressure lift, with buoyancy contributing considerably less [5]; this is in stark contrast to a displacement hull, which is mainly supported by buoyancy. The main characteristic of a planing hull is that it provides for flow separation at the transom and along the sides. While there is no definitive quantification of when a hull begins to plane, the general guidelines is that at speeds above a length Froude Number of 1.0 to 1.2, clean flow separation from the chines and transom is achieved for the inception of planing [6]. For an extended period, the planing hull was dismissed as a design, as it could not attain its promised advantages in rough seas, until the re-evaluation which shown drastic improvements in seakeeping characteristics with proper selection of hull form proportions and loadings [7].

2.1.2 RESISTANCE PERFORMANCE

Wherein the design of a planing craft was motivated by a desire to reduce wave-making resistance, the eureka moment of the development of advance hull forms was born from an idea to lower the viscous resistance. By incorporating innovative design, optimisation techniques and combining principals of hydrodynamics, ST Engineering is able to provide a wide range of different hull forms capable of achieving incredible speeds.

Methods such as creating a step at which the flow separates, the aft part of the hull (with low hydrodynamic pressures) can be ventilated, reducing wetted surface area while still maintaining the high hydrodynamic pressure forces necessary in the forward part of the hull, to ensure sufficient lift [5] modifying the chine to overcome bow drop and porpoising effect and creating a layer of air to reduce the resistance of the hull.

2.1.3 INSUFFICIENT CAPABILITIES

It has always been a challenge for naval architects to balance speed, length and displacement performance for high-speed crafts as they are closely related. The easiest way to achieve a low displacement is to limit the length, amount of payloads and deadweight carried on board. With lesser room to cater for self-defence and deterrence and support capabilities such as search and rescue, conventional high speed crafts are too one-dimensional and not flexible enough to respond to the ever-changing and unpredictable threats.

Speed should not be the only salient feature when it comes to high-speed craft designs. Modern designs are to be highly versatile in managing these situations swiftly and promptly.

2.2 POOR FOCUS ON HUMAN FACTORS

Designers place a lot of emphasis on achieving remarkable speeds in ensuring aspects such as range, seakeeping and concept of operations (CONOPS) are well covered and validated for high-speed crafts but human factor considerations are often neglected and often one of the least priorities.

Crew suffers the most as their nature of work requires them to work long hours on board such high-speed crafts. Injury due to high vertical accelerations, constant exposure to noise levels beyond permissible exposure limit (PEL) and uncomfortable working environment due to high vibration levels will led to long-term side effects.

Human factors and ergonomics are equally important because the crew are the most important asset on board. The success of the operation is dependent on the crew's wellbeing.

2.3 LACK OF OPERATIONAL FOCUS

Operators and designers often have different views in terms of design aspects. Designers are generally objective driven while operators tend to look at finer details to better facilitate and to streamline their daily workflow. Very often, such differences lead to several rounds of design iterations and will influence the final design.

The importance of understanding the operator's needs and striking a balance will be the key to develop a well-balanced design to avoid complications and to achieve the state-of-the-art design.

3. WINNING FORMULA

3.1 SUPER SWIFT SERIES

The development of the Super Swift series, as shown in Figure 2, is the solution offered by ST Engineering in enhancing coastal surveillance and security.

Upon identifying the problems and shortfalls of conventional high-speed crafts, ST Engineering aims to overcome these limitations. The designers from both Singapore and US came together to brainstormed for new ideas, share valuable experiences from past projects, gathered feedbacks from potential customers to understand their needs and requirements. These information are the core foundation in building up the design of the Super Swift series fast patrol boat or interceptor.

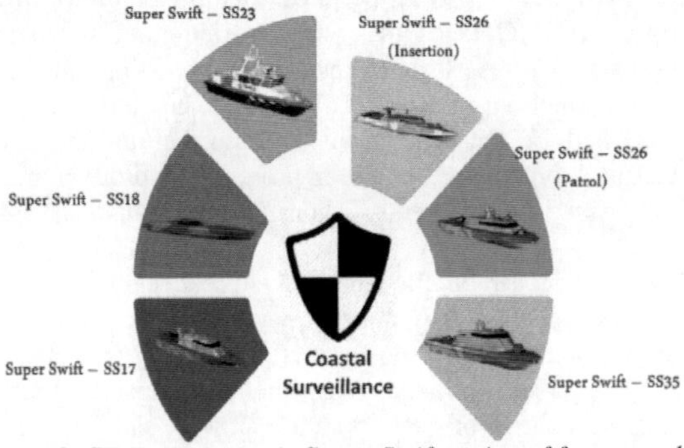

Fig. 2. ST Engineering's Super Swift series of fast patrol boat

3.2 DESIGN ACHKNOWLEDGEMENT

ST Engineering design efforts in the development of the Super Swift series were acknowledged when we were awarded with the contract to design, construction and maintain a series of fast patrol boat based on the Super Swift 23 design towards near 2018. The new series of patrol boats will replace the existing older patrol boats so as to strengthen its capabilities against maritime threats which is meant for ultra-high-speed operations of up to 60 knots, with enhanced manoeuvrability and crew comfort.

To overcome the conventional design methods, which were often deemed as unsuitable due to the numerous sensitive design parameters such as weight control, speed-powering performance and seaworthiness, ST Engineering introduced various SMART initiatives through thorough understanding of the operating needs from past experience, in-depth research and detailed studies before introducing the design for selection.

These SMART initiatives form the foundation for all the Super Swift series of fast patrol boats.

3.3 SMART INITIATIVES

3.3.1 SMART HULL FORM

A wide variety of hull forms were developed for the purpose oft the Super Swift series.

The baseline design original hull (OH) form used for the development of the Super Swift series was derived from well-proven 17m Deep Vee planing hull as shown in Figure 3, 4 and 5 used for a luxury yacht back in the early 1990s.

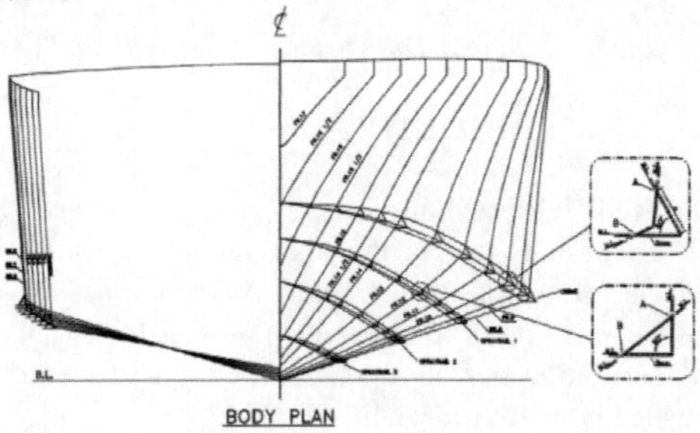

Fig. 3. Body Plan of the Original Hull (OH)

Fig. 4. Scaled Model of Original Hull (OH) used for Model Test

Fig. 5. 3D illustration of the Original Hull (OH)

Based on the well-proven hull form, the following hull forms were developed in-house through intensive studies conducted through Computational Fluid Dynamics (CFD) simulations and validation through model test:

- Step Hull (SH);
- Air-Cavity Hull (ACH);
- Enhanced Hard Chine Hull (EPH);

The parameters of the various hull form shown in Table 6.

	OH	SH	ACH	EPH
L_{OA} [m]	17.00	17.00	17.59	17.00
L_{PP} [m]	16.54	16.54	16.54	16.54
B [m]	4.32	4.32	4.32	4.32
T_F [m]	0.68	0.65	0.61	0.68
T_A [m]	0.78	0.74	0.97	0.78
∇ [m^3]	21.46	21.46	21.46	21.46
Δ [t]	22.00	22.00	22.00	22.00
L_{CG} [m]	5.71	5.71	5.71	5.71

Table 6. Principal Dimensions of SMART Hull forms

3.3.1.1 STEP HULL (SH)

The Step Hull was introduced to generate low pressure at the aft of the step as the craft moves forward, which introduces suction in air from the sides. Increase in speed will generate more lift and which will enable to hull to be free of the water. Friction and drag will be reduced drastically hence reduce in resistance.

As shown in Figure 7, 8 and 9, the step was carefully positioned at the aft of the near the LCG to allow optimal suction in air from the sides when the vessel is accelerating. Such a modification has improved the resistance by approximately 5 % as confirmed and validated in scale-down model tank test.

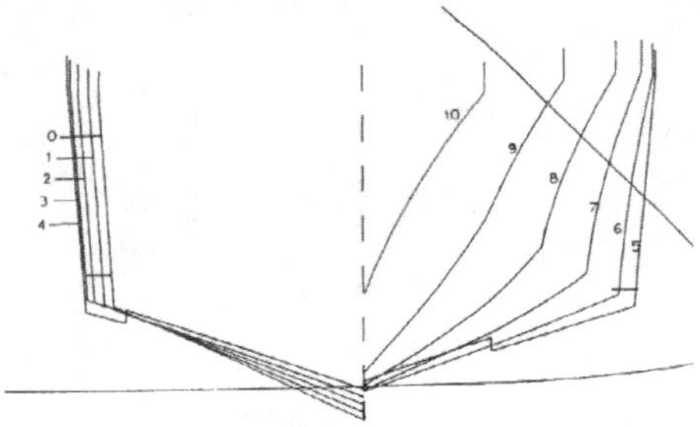

Fig.7. Body Plan of Step Hull (SH)

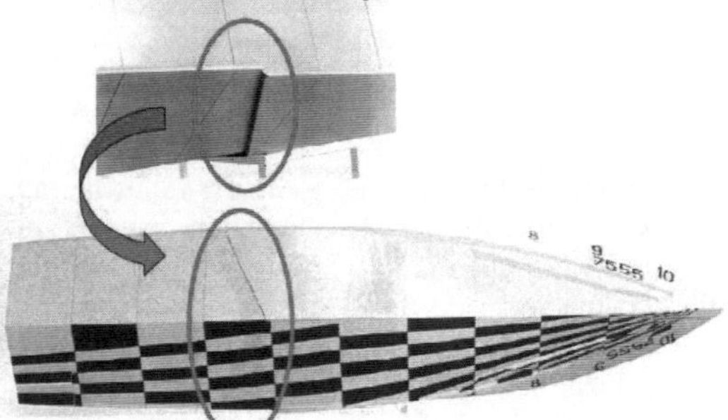

Fig. 8. Scaled Model of Step Hull (SH) used for Model Test

Fig. 9. 3D illustration of the Step Hull (SH)

The principle of a low pressure aft has proven to provide favourable returns as compared to her initial basis hull form.

3.3.1.2 AIR CAVITY HULL (ACH)

Air has a lower frictional resistance than water. By creating a layer of air cavity between the hull and the water surface, the hull resistance could be reduced. The Air Cavity Hull (ACH) applies this principle to the Original Hull (OH) to further reduce frictional resistance and improve vessel performance. By creating an air cavity at the bottom of the hull form as shown in Figure 10, 11 and 12, the layer of air between the hull and the water cushions the vessel and improves its hydrodynamic performance as validated by rigorous seakeeping tests.

Comprehensive model testing showed that the ACH was able to reduce drag by approximately 12% to 17% for waterjets and 17% to 20% for surface piercing propellers in the high-speed range. This directly translates into higher speed gain of 4 to 4.5 knots using the same engine and for the same displacement. The ACH will be ideal hull form for vessels that require to beyond the speed of 60 knots with the flexibility to consider both waterjets and surface piercing propellers.

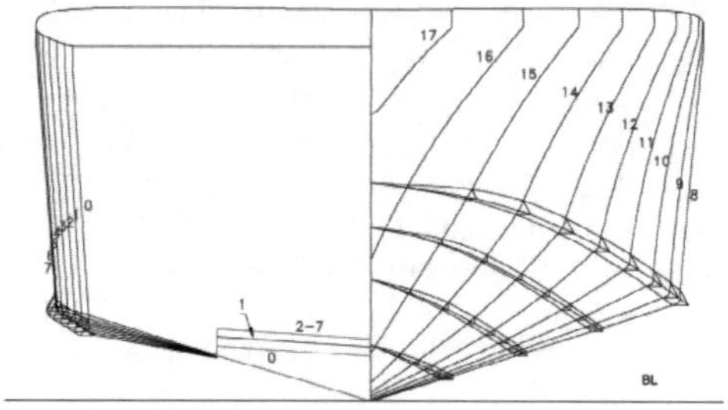

Fig. 10. Body Plan of Air Cavity Hull

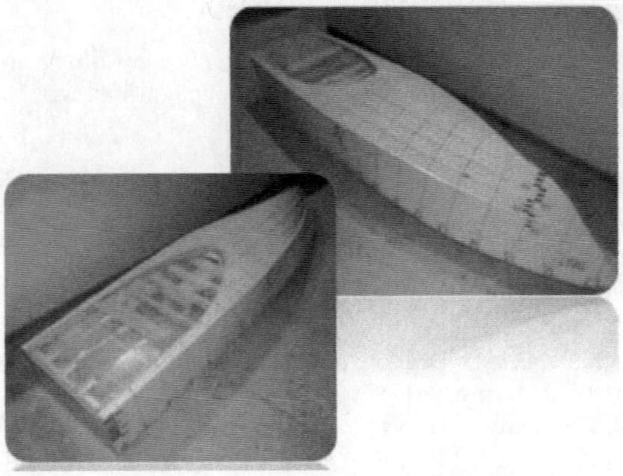

Fig. 11.. Scaled Model of Air Cavity Hull (ACH) used for Model Test

Fig. 12. 3D illustration of the Air Cavity Hull (ACH)

3.3.1.3 ENHANCED PLANING HULL (EPH);

The Enhanced Planing Hull (EPH) sets a different focus and goal as compared to both the SH and ACH. The purpose was to provide a simple solution to overcome the shortfall of the Original Hull (OH). The intention was to refine and improve the initial OH with the least amount of changes and yet achieve favourable returns.

A series of Computational Fluid Dynamics (CFD) analysis were conducted to understand the OH and it was identified that there is an obvious amount of low-pressure area grows and intensifies when the vessel is travelling at a speed of 65 knots, as seen in Figure 13. This indicates a possible tendency to dynamic instability when travelling at high speed.

Fig. 13. Low-pressure distribution at the aft of the Original Hull (OH) at 65 knots

Dynamic instability is extremely dangerous, especially for high-speed-vessels, and it is to be avoided at all cost. Several steps were taken by considered bringing the transom down to the baseline, making the keel line flat along the baseline. The hull lines produced by doing this can be seen below in Figure 14.

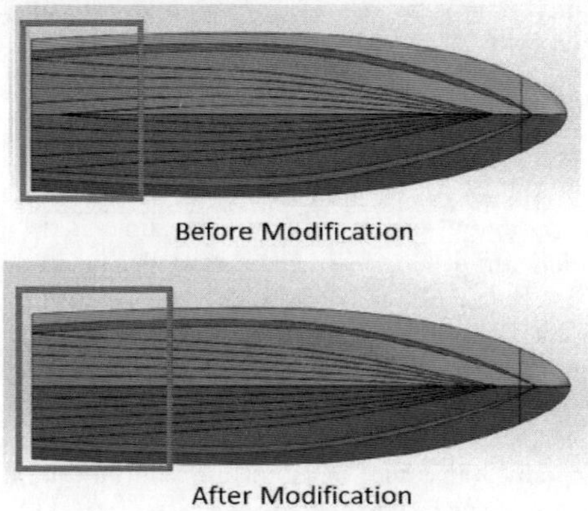

Fig. 14. EPH Optimisation

Such a modification has significantly reduced the low-pressure point at high speeds and has proven to be very effective as shown in Figure 15.

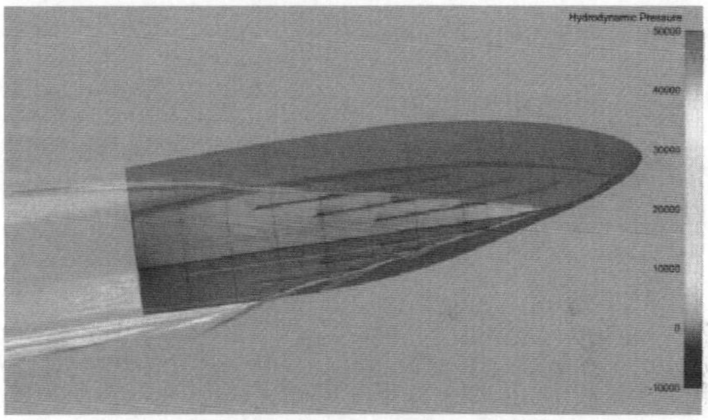

Fig. 15. Pressure distribution at the aft of the Enhanced Planing Hull (EPH) at 65 knots

The performance of the EPH was further validated during the model test and has proven to correspond very well with the results obtained from the CFD analysis.

3.3.1.4 HULL FORM COMPARISON

All of the above SMART hull forms developed were validated using both CFD analysis and model testing.

Detailed resistance to speed comparison study will be conducted based on the operational requirements prior to us recommending the best suited hull form to the customer. The Super Swift series offers wide flexibility when it comes to the selection of the hull form.

3.3.2 CREW FOCUSED

Conventional design places too much focus and emphasis on the operational requirements, resulting in crew comfort and/or ergonomics becoming a lower priority. The crews onboard are the most important asset as they are the key users when it comes to the day-to-day operations. Thus. Crew comfort and safety should be considered as equal importance, when it comes to design requirements.

High-speed craft operations often result in physical and mental fatigue [8], resulting in extreme discomfort and unwanted injury leading to crew downtime. It is also important for designers to understand that apart from fatigue, the access to periods of rest is equally as important [8]. By gathering feedback from our clients based on past experience, we are able to better understand their operational requirements. Such feedback has enabled the designers to identify the key critical areas and to ensure that the limits set are safe and negative effects of fatigue can be limited.

The Super Swift series looks into all the key human factors and ergonomics especially when the design of the Super Swift series is meant to operate at speeds of up to 60 knots.

3.3.2.1 QUIET DESIGN

Crews on board will spend most of their time in two areas – bridge and accommodation. It is important for these two areas to be as quiet as possible.

The bridge area is where all the navigation and communication take place. Crews will be required to respond to every single command and/or distress signal within seconds. If the bridge design is not well considered, it will disrupt key messages, which will influence both the decision-making process and the operation of the vessel.

Crew accommodation is where the crew will take their short intervals of breaks when they are operating out at sea. Those pockets of time are important for them to rest and keep themselves alert when they are on watch.

The aim and objective is to achieve low noise levels of not exceeding 75dB (A) at high speeds. Such an objective seems like an easy task by simply adding more acoustic insulation but the weight impact due to the acoustic insulation will influence the speed of the vessel. ST Engineering explored several possible methods to further reduce the noise levels with the least weight impact as possible.

The following methods are highly flexible and can be used at either compartment depending on their design requirements.

3.3.2.1.1 FULLY FLOATING CABIN

Suspended deckhouse concept is achieved by fitting an isolated deckhouse on the resilient mounts, which will be installed on the deck

Such a construction is where the floor, walls and ceiling form a "box" which is vibration isolated from the vessel structure. Acoustic insulation is used to act as a "spring" to detach/mount the cabin floors, walls and ceilings.

Such a concept will allow a significant noise reduction of up to approximately 8 to 10 dB (A). The estimated weight increase is approximately 1% of the entire displacement of the vessel.

3.3.2.1.2 SEMI-FLOATING CABIN

An alternative to a fully-floating cabin will be a semi-floating cabin, it is able to reduce the weight increase by 50% of what is required for the fully-floating cabin and still provide noise reduction of up to approximately 4 to 5 dB (A). As illustrated in Figure 16, the semi-floating floor, interior fit out is constructed using a fully floating floor with partial floating walls, single glazed windows and a resiliently mounted sheet metal ceiling.

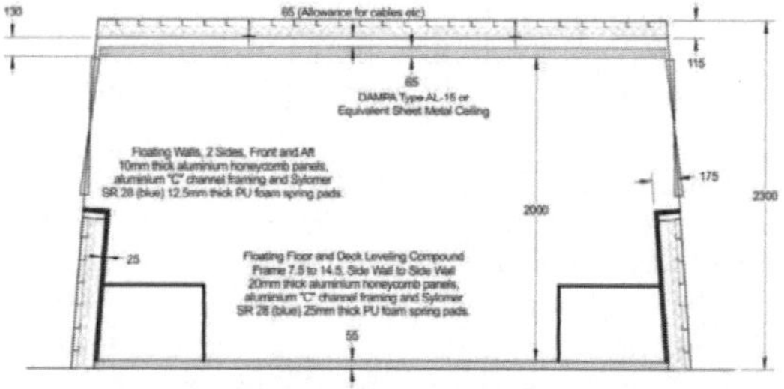

Fig. 16.. Sketch of the semi-floating floor concept

It is important for naval architects to look into different options depending on how much weight margin is available. Both proposals will offer excellent noise reduction for both the bridge and accommodation area.

3.3.2.2 LOW VERTICAL ACCELERATION

Conventional high-speed hull forms are not well optimised and tend to be subjected to high levels of vertical acceleration. The MK5 was reported to be subject to extremely high vertical accelerations and it has caused extreme impact to the crew onboard the MK5 [4].

Such high vertical acceleration will cause crew injury and fatigue, which will affect the crew's daily operation.

As part of our SMART hull form optimisation process, all our hull forms were carefully analysed to overcome such an unfavourable vertical acceleration phenomenon. The Super Swift series designs are able to achieve low vertical acceleration values of less than 1G at speeds up to 40 knots as tested and verified by the model test as shown in Figure 17.

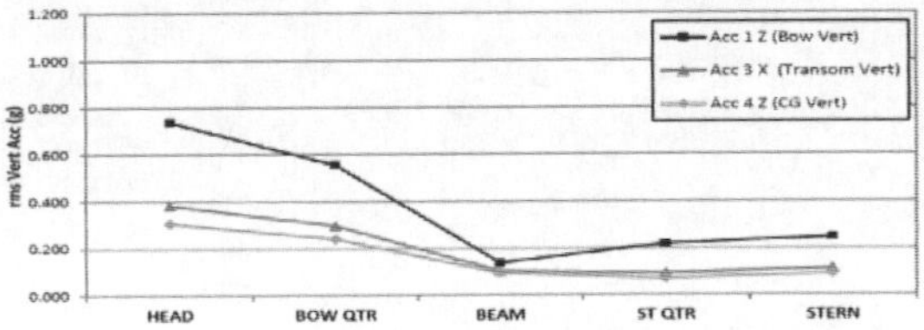

Fig. 17. Less than 1G Vertical Acceleration performance at 40 knots

3.3.2.3 ERGONOMICALLY OPTIMISED LAYOUT

Conventional designs layouts tend to be very compact and hinders the workflow of the crew. Many incidents were also reported due to compact designs [5].

All the layouts on board the Super Swift series were carefully considered to ensure a head room of more than two (2) meters in height, with minimum eight hundred (800) millimetres of passageway width. Such headroom height and passageway width are important to ensure that the crew have sufficient room to move from one point to another as part of their daily operational requirement.

Controls and monitors are well arranged to cater to each crew's task on board the vessel. It is important to ensure that the crew is able to perform task seamlessly and simultaneously in the event if there is a need to perform dual roles.

Shock-absorbing seats were considered for all the crew on board to ensure that all possibilities of any unwanted high vertical acceleration would be eliminated when the crew is operating the vessel.

3.3.3 MUTLI-ROLES CAPABILITY

Due to the size and space limitation, conventional patrol boat of sizes ranged between 17m to 35m typically lack multi-role capabilities and tends to be very singular objective focus.

It is important for modern patrol boat to be able to perform multiple roles. The Super Swift series can be configured to perform several roles such as firefighting, search and rescue and even carry various small crafts via ST Engineering's innovate unique Quick Launch and Recovery System (Q-LARS).

3.3.3.1 QUICK LAUNCH AND RECOVERY SYSTEM (Q-LARS)

ST Engineering developed our in-house Q-LARS which allows enables various small crafts such as Unmanned Surface Vehicle (USV) and Rigid Hull Inflatable Boat (RHIB) to be launched and recovered quickly without the need of using a conventional winch system.

The Q-LARS as shown in Figure 18. is scalable and suitable for a wide range of small crafts, which includes a stern ramp door, slipway, wheel sets, winch and hydraulic system.

The system can be operated up to sea state 3, and the foundations, fittings and equipment can withstand wind loading, vibration and dynamic forces produced by motion in sea state 7 and requires only 2.5 minutes launching and recovering a small crafts as compared to a conventional system whereby it requires more than 10 minutes.

Such a system can easily be configured on board the Super Swift Series to enable her to perform simultaneous operations.

Fig. 18. Artist impression of the Quick Launch and Recovery System (Q-LARS)

4. CONCLUSIONS

The Super Swift series patrol boat are the result of a carefully considered design. By understanding the shortfalls of existing conventional designs, it has given ST Engineering an edge when it comes to such high-speed designs.

The highly flexible and configurable design allows it to meet a wide range of operating requirements. With a range of sizes between 17m to 35m, ST Engineering is confident that the Super Swift series is able to meet all the stringent requirements and expectations from the operators such as coast guards.

5. REFERENCES

1. Recaap isc. 2019. Piracy And Armed Robbery Against Ships In Asia. Recaap Isc Annual Report 2019: Page 5, 8.
2. Dand, I W, (2006). High Speed Craft Bow Diving In Following Seas. In High Speed Craft - ACV's, WIG's & Hydrofoils. London, 31-Oct-2006 to 1-Nov-2006.
3. 'Guidance to the Master for Avoiding Dangerous Situations in Following and Quartering Seas'. MSC/Circ. 707, IMO London, October 1995
4. Ensign, W., 2000. A Survey Of Self-Reported Injuries Among Special Boat Operators. Naval Health Research Center.. Report No. 00-48, 4.
5. Faltinsen, Odd. 2005. Hydrodynamics of High-Speed Marine Vehicles, New York, Cambridge University Press
6. Savitsky, Daniel. 1992. Overview of Planing Hull Developments. Proceedings of the Intersociety High Performance Marine Vehicle Conference and Exhibit: HPMV'92. Arlington, VA. American Society of Naval Engineers.
7. Savitsky, Daniel; Gore, Jerry L. 1980. Re-Evaluation of the Planing Hull Form. Journal of Hydronautics. 34-47.
8. A. W. S. Leung, C. C. H. Chan, J. J. M. Ng, and P. C. C. Wong. Factors contributing to officers' fatigue in high-speed maritime craft operations. Applied Ergonomics, 37:565–576, 2006.

6. AUTHORS BIOGRAPHY

Wen Hao Kenneth Lee is holding the position of Assistant Principal Engineer in ST Engineering Marine Ltd. He graduated with a B.Eng. (First Class Honours) in Naval Architecture in 2014 from Newcastle University upon Tyne. He was one of the designers responsible in the 25m Specialised Marine Craft for the Republic of Singapore Navy and the latest edition would be 23m 5th Generation Patrol Boat for the Singapore Police Coast Guard.

AN INVESTIGATION INTO THE RESISTANCE CHARACTERISTICS OF THREE-STEP PLANING HULLFORMS

J. Fan and D. G. Trodden, Newcastle University, UK

SUMMARY

Existing research into stepped planing hulls have focused on the effects of step parameters, but there appears to be a lack of research into the effect of the number of steps beyond two. This investigation examines the effect on resistance of implementing three steps for a planing hull through a two-pronged approach: performing experimental tests and running CFD simulations. The basis hull used is the Series 62 model, with modifications are to create a three-step variant based on existing literature on optimum stepped hulls to compute flow separation reattachment length. Cross-examination of experimental and numerical data for the stepless model show excellent agreement, validating the CFD model used. Results from the study suggest that further optimisation of step design can produce performance gains.

NOMENCLATURE

The following symbols, abbreviations, and acronyms are used throughout this paper:

Fn_∇	Volumetric Froude number (-)
V	Ship Speed (m s^{-1})
g	Gravitational Acceleration (m s^{-2})
Δ	Mass of Displacement (kg)
∇	Volumetric Displacement (m^3)
β	Deadrise Angle (°)
β_1	Deadrise Angle of Forebody Planing Surface (°)
β_2	Deadrise Angle of Aftbody Planing Surface (°)
$\triangleleft$	Angle between Forebody Keel & Aftbody Keel (°)
ε	Inclination of Thrust Line relative to Keel Line (°)
e	Distance below keel where propeller shaft passes (m)
υ	Kinematic viscosity (N s m^{-2})
ρ	Density of water (kg m^{-3})
P	Pressure (N m^{-2})
0S	Stepless Variant
3S	Three-step Variant
B_1	Beam of Forebody Planing Surface (m)
B_2	Beam of Aftbody Planing Surface (m)
B_{chine}	Chine Beam (m)
B_{max}	Maximum Chine Beam (m)
CAD	Computer Aided Design
CFD	Computational Fluid Dynamics
G_0G_1	Shift in LCG (m)
G_1G_2	Shift in VCG (m)
L_{K2}	Afterbody Wetted Keel Length (m)
L_{PP}	Length between Perpendiculars (m)
LS_1	First Step Position from transom (mm)
LS_2	Second Step Position fr. transom (mm)
LS_3	Third Step Position fr. transom (mm)
LCG	Longitudinal Centre of Gravity measured forward of transom (m)
R_P	Hull Pressure Resistance (N)
R_T	Total Hull Resistance (N)
R_V	Hull Viscous Resistance (N)
TC	Test Configuration
W	Vessel Weight (N)
WSA	Wetted Surface Area (m^2)
VCG_1	VCG Position above Keel (m)
VS	Step Height in millimetres (mm)
VS_{chine}	Step Height as a ratio of chine beam (-)

1. INTRODUCTION

The aim of this study is to examine the effect of additional steps – beyond the conventional one and two-step hulls – on the resistance characteristics of a planing craft. To achieve this, a three stepped variant of a Series 62 hull form is created and tested at planing speeds in varying conditions, whereby the resistance is measured and compared. To this end, both numerical and experimental methods are employed to obtain and verify the required data in this investigation.

Planing hulls were first introduced as a means to overcome the issue of a main resistance hump common to displacement vessels. Stepped hulls are an innovation on planing hulls to further bring down resistance. Previous numerical research have focused on single-step hulls, but there is a lack of experimental research and a lack of examination into the effects of additional steps beyond a number of two.

The main aim of this study is to establish if resistance benefits are to be gained over a conventional planing hull through the use of a three-step hull.

2. BACKGROUND THEORY

2.1 STEPPED HULLS

The stepped hull achieves resistance improvement over a conventional planing hull by seeking a reduction in WSA. This is possible through the separation of water from the hull aft of the step. Previous research has established that the aft portion of the hull makes a negligible contribution to planing lift due to the low water pressure in this area, and yet plays a significant role in frictional resistance [1]. The introduction of a transverse step has been proven to be useful in eliminating ineffective wetted area in the aft region by forcing water flowing along the bottom to separate from the hull at the step, with the steps becoming useful over a stepless hull above Fn_∇=2.68 [2].

2.2 STEP HEIGHT

It has been shown that below Fn_∇=2.7, resistance increases with step height because the vessel does not travel at high enough speeds for proper flow separation [2]. Above this critical speed, the reattachment length increases linearly with step height, decreasing the WSA and hence reducing frictional resistance [3]. However, an increasing step height also results in in a larger trim angle which corresponds to added wavemaking resistance [4]. This can decrease the benefits to be gained from a reduced frictional drag.

One study concluded that a step height corresponding to 4% of the chine beam offers the optimum compromise [2], with another proposing that the optimum range for step height lies between 0.05 to 0.06 of the beam [3].

2.3 SVAHN'S STEPPED HULL METHOD

Svahn combines Savitsky's original planing hull method [5] with the equations for determining afterbody hull shape relative to the wake surface [6] to develop a semi-empirical method for predicting the performance of a stepped planing hull [7]. Comparison of calculations and full-scale experimental data shows a promising correlation.

2.4 NON-DIMENSIONAL SPEED COEFFICIENT

With planing vessels, the volumetric Froude Number (as defined in Equation 1) should be chosen as the non-dimensional parameter for comparison of performance between vessels to ensure the same full-scale speed amongst different designs [8].

$$Fn_\nabla = \frac{V}{\sqrt{g\nabla^{\frac{1}{3}}}} \tag{1}$$

3. METHODOLOGY

3.1 BASIS PHYSICAL HULL FORM

An existing model hull of a Series 62 4667-1 previously employed in a separate investigation at the Newcastle University Hydrodynamic Laboratories was used for this investigation. Experimental data from the original towing tank tests of the stepless basis model [9] were also obtained to reduce the number of tests necessary[1].

3.1 (a) Original Experimental Test Runs

A total of four test configurations were set up for the original towing tank tests of the stepless 4667-1 hull form to collect resistance data for varying LCG and loading configurations. Test runs were conducted for speeds ranging from a Fn_∇ of 0.50 up to the maximum possible Fn_∇ at the highest possible test speed[2]. The key parameters of each model configuration are shown in Table 1.

Table 1: Summary of Stepless Model Test Configuration

	TCs			
	0S-TC1	0S-TC2	0S-TC3	0S-TC4
L_{PP} [m]	1.00			
B_{max} [m]	0.244			
β [deg]	12.5			
Δ [kg]	6.92	6.92	4.83	4.83
LCG [m]	0.448	0.408	0.448	0.408
Maximum Fn_∇ [-]	2.75	2.75	3.00	3.00

3.2 ADAPTING SVAHN'S METHOD

As outlined in Section 2.3, Svahn presents a method for power prediction of a single-step planing hull. This is an iterative method that splits the hull at the step into a forward hull and an aft hull to seek an equilibrium of vertical forces. When a vertical equilibrium has been achieved, a moment equilibrium is then sought to determine the resistance of the hull. In seeking this balance, the trim angle, resistance, and flow separation length can be determined.

3.2 (a) Program Development

To aid the process of analysis, a Python script implementing Svahn's Method is written. The algorithm for this is illustrated in Figure 1.

3.3 STEP HEIGHT DETERMINATION

To determine the step height of the three-step variants, a flow separation investigation using Svahn's Method is performed.

Due to a lack of empirical or semi-empirical power prediction methods for multi-step hulls, this investigation employs Svahn's method for predicting the flow separation at each step for multi-step hulls. A parametric study for multi-step models is conducted to determine the

[1] While it is ideal to conduct a full set of test runs again – as the model conditions may have changed – it is not practical to do so due to the limited time frame in the tow tank available for experimental testing, and thus reusing the experimental data from original test runs is an acceptable compromise.

[2] The highest possible test speed is the maximum attainable carriage speed in the Newcastle University Hydrodynamics Laboratory Towing Tank – dependent on factors to be explained in Section 3.7.2

optimum step parameters for a three-step variant based on the following constraints:

- Position of first step maintained at 0.308m forward of transom
- Achieving maximum flow separation length[3]
- Based on a maximum running speed corresponding to $Fn_\nabla = 2.75$

The key stages of this study are shown in Figure 2, while Figure 3 illustrates the resulting parameters from each stage.

3.3 (a) Stage 1

The maximum available length for flow separation is defined as the total length of hull aft of the first step.

3.3 (b) Stage 2

A target reattachment length (flow-separation length) for each step is arbitrarily set at a suitable value that would allow separation from all steps to re-join before each subsequent step, and before the transom of the hull. It is assumed that the flow separation length at each step remains constant[4].

3.3 (c) Stage 3

A buffer is inserted between the reattachment point of the flow and the position of the next step. Intended to allow for limited error, the gap provides confidence that the flow will re-join the hull prior to the subsequent step – even if the actual separation length overshoots the predicted length – thus ensuring that flow separation can continue to occur at each step.

This is important because the vessel's trim angle would increase substantially should the flow miss its reattachment point for a step, producing higher pressure resistance while failing to achieve the maximum possible reduction in wetted surface area (and hence viscous resistance). The overall effect of a failure for flow to re-join before the end of a step is likely to be an increase in resistance over a basis stepless planing hull.

3.3 (d) Stage 4

The target reattachment length is now subtracted from the total flow reattachment length to obtain a value of L_{K2} for the first step[5].

Key flow reattachment parameters for a full-scale three-step hull[6] from Stage 1 through Stage 4 are summarised in Table 2.

Table 2: Summary of Key Parameters from Flow Separation Analysis Stages 1-4

Stage	Parameter	Value
(1)	Maximum Available Length for Flow Reattachment [m]	5.544
(2)	Target Reattachment Length for Individual Step [m]	1.5
(3)	Buffer Gap between Reattachment & Next Step [m]	0.35
(4)	Target L_{K2} for First Step [m]	4.044

3.3 (e) Stage 5

By feeding the full scale vessel parameters (shown in Table 3) into the developed program for applying Svahn's Method, multiple runs are carried out for different step heights ranging from 1% to 6% of the B_{chine} to determine the L_{K2} for each variation. Table 4 shows the full-scale results for varying step heights.

Table 3: Input Parameters for Flow Separation Analysis

	Full-Scale	Model
Δ [kg]	41366	6.92
β_1 [deg]	12.5	12.5
β_2 [deg]	12.5	12.5
Φ [deg]	0	0
ε [m]	10	10
e [m]	0	0
LCG_2 [m]	7.344	0.408
VCG_1 [m]	1.186	0.066
B_1 [m]	4.392	0.244
B_2 [m]	4.392	0.244

Table 4: Variation of Afterbody Wetted Keel Length with Step Height

[3] Maximum flow separation length is defined as the maximum possible length of flow separation at each step such that the flow re-joins the hull prior to the commencement of the next step – i.e. the flow having separated from the first step (foremost step) should reattach to the hull a short distance before (forward) of the second step.

[4] In reality this is unlikely to be true since the disturbed flow re-joining the hull prior to each step would have a significantly different angle of intersection and thus a wake profile different from the one used in the semi-empirical methods of Svahn. However, for the scope of this analysis and considering the lack of prior research, this is a reasonable simplification to make which can then be adjusted for further work based on the results from this study.

[5] This value assumes the entire hull aft of the first step to be one hull and neglects the presence of additional steps which is irrelevant in analysing the reattachment point of the forward most step.

[6] Due to the limitations of Svahn's method, analysis for flow separation is carried out at full scale. The scaling ratio of full scale to model is 18.

VS_{chine} [-]	0.06	0.05	0.04	0.03	0.02	0.01
VS [m]	0.264	0.220	0.176	0.132	0.088	0.044
L_{K2} [m]	2.812	3.115	3.445	3.810	4.244	4.718

3.3 (f) Stage 6

Interpolating between data points developed in Stage 5 to seek a match with the target L_{K2} defined in Stage 4, the theoretical optimum step height for a vessel operating under the constraints listed earlier can then be determined. The final required VS_{chine} is established to be 0.024. This translates to a VS_{model} of 6.0mm.

3.4 CREATION OF CAD MODELS

Using hull lines obtained from the original experiments [9], a CAD model of the original planing hull is created in RHINO.

Using the step heights determined by the process in Section 3.3, the original hull lines are modified to produce a three-step variant. The steps are set to be of equal height and positioned to have an equal length distribution.

Key parameters of the three-step variant are summarised as follow:

- VS = 6.0 mm
- LS_1 = 308.0 mm
- LS_2 = 205.1 mm
- LS_3 = 102.4 mm

Figure 4 shows a visualisation of the three-step hullform.

3.5 PHYSICAL MODEL MANUFACTURE

The hull lines for the three-step variant were supplied to the technicians at the Newcastle University Hydrodynamics Laboratory, whereupon three steps were cut into the original 4667-1 stepless model hull according to specifications. Waterlines corresponding to two different loading conditions were then drawn for easy positioning of the vessel in water as shown in Figure 5.

3.6 DETERMINATION OF PARAMETERS FOR STEPPED VARIANT

3.6 (a) Mass of Displacement

The original hull was tested at two different masses of displacement: 6.92kg and 4.83kg. This investigation will aim to reproduce an equivalent condition, after having accounted for the steps, to ensure continuity of results. To determine the new mass of displacements for the stepped variants, a simple volumetric scaling is performed using Equation (2):

$$\rho = \frac{\Delta}{\text{Hull Volume}} \tag{2}$$

Assuming a constant hull density, the volume removed from the hull is found in RHINO, thus allowing for the new mass of displacement to be computed using a ratio of new hull volume to original hull volume as summarised in Table 5:

Table 5: Mass of Displacement Calculations

	0S	3S
Step #1 Removed Volume [m^3]	-	0.0001221
Step #2 Removed Volume [m^3]	-	0.0002250
Step #3 Removed Volume [m^3]	-	0.0003057
Total Removed Volume [m^3]	-	0.0006528
∇ [m^3]	0.02712	0.02647
$\Delta_{TC1,TC2}$ [kg]	6.92	6.75
$\Delta_{TC3,TC4}$ [kg]	4.83	4.71

3.6 (b) Draft Determination

Using the new mass of displacements, the required draft for a level trim is then determined through an analysis in MaxSurf. This is established to be 92.5mm and 82.5mm from deck for the heavier and lighter loading configurations respectively.

3.6 (c) Centre of Gravity

To determine the new centre of gravities for the stepped variants, a moment equilibrium of the removed steps is sought about the original centre of gravity:

$$\text{Shift in CG} = \frac{\text{Moment of Step(s) Volume}}{\text{Volume of Removed Step(s)}} \tag{3}$$

Table 6: LCG & VCG of Three-step Variant

		0S	3S
	G_1G_2 [m]	-	0.00147
TC1, TC3	G_0G_1 [m]	-	0.00795
TC1, TC3	LCG [m]	0.448	0.45595
TC1, TC3	VCG [m]	0.0763	0.0778
TC2, TC4	G_0G_1 [m]	-	0.00696
TC2, TC4	LCG [m]	0.408	0.41496
TC2, TC4	VCG [m]	0.0763	0.0778

By measuring the centroid of each removed step in RHINO, and using the volume of each removed step from Section 3.6.1, the moment generated by each step can be determined. The shift in both longitudinal centre of gravity and vertical centre of gravity can thus be determined using Equation (3). The final centres can be determined by summing up the shifts caused by each individual step and factoring the total shift into the original stepless hull's LCG & VCG. This is shown in Table 6.

3.7 EXPERIMENTAL TESTING

The three-step physical model was tested in the Newcastle University Hydrodynamics Laboratory towing tank to obtain resistance values for comparison with the original planing hull.

3.7 (a) Newcastle University Towing Tank

Specifications of the tow tank are listed below [10]:

- Tank Length: 37 m
- Tank Width: 3.7 m
- Normal Water Depth: 1.25 m
- Maximum Normal Carriage Speed: 3.5 ms^{-1}

3.7 (b) Experimental Test Matrix

As the limited tow tank time available for testing inhibits the possibility of re-testing the original hull before cutting steps into it and testing the modified hull, the test matrix is designed to mirror the parameters in the original test runs by Bowles in 2001.The model is tested at varying speeds with different masses of displacement. The maximum Fn_∇ is determined by allowing for an adequate testing length to collect sufficient data while catering for ample braking distance before the end of the tank. Given that the maximum shift in LCG is approximately 7.95mm – less than 1% of the model length – the LCG shift is considered insignificant and the heave post is positioned on the original LCG[7].

A summary of the test matrix is presented in Table 7.

Table 7: Test Matrix

	Δ [kg]	Draft [mm]	LCG [m]	Fn_∇ [-]
3S-TC1	6.75	51.5	448	0.50 – 2.75
3S-TC3	4.71	42.5	448	0.50 – 3.00

3.7 (c) Model & Experimental Set-up

In simulating the experimental set-up of the original experiments by Bowles, the model was set up on the towing carriage with the freedom to heave and trim. The towing point (dynamometer attachment point) is kept the same as that for the original model. While not ideal, this avoids the introduction of an additional variable into the test matrix, allowing for the three-step bare-hull test results to be directly compared to the original stepless results.

During each test run, data acquisition software measures resistance (drag), heave, pitch, and velocity at a hundred samples per second.

As all test run conditions fall the Reynolds Number of $5x10^6$ required for fully turbulent conditions, a turbulence stimulating strut is added to trip up the flow and ensure that the model is tested in turbulent flow conditions. The strut is positioned between one to two times of the maximum vessel beam forward of the bow.

3.7 (d) Counter-balancing & ballasting

The total combined weight of the towing set-up, including the heave post, is 4.896kg. The model itself weighs 4.504kg. Since the vessel does not have sufficient displacement to absorb the weight of the dynamometer, a counter-balancing system is required to take the weight of the towing block system off the model and bring it back to the required mass of displacement. The counter-weights will also need to absorb the ballast weights used to trim the model.

To this end, a string is attached to the heave post and run through a pulley system leading to a weight base plate positioned near the aft end of the towing carriage. Weights are then incrementally added until the desired waterline is achieved. To minimise the force acting on the vessel arising from acceleration of mass as the weights swing around during carriage movement, additional strings linking the base plate and the carriage posts are employed to restrict weight movement (circled in Figure 6).

Table 8: Associated Weights for Experimental TCs

	Ballast Weight [kg]		Counter Balance Weight [kg]
	Fwd	Aft	
3S-TC1	0.950	1.000	4.623
3S-TC3	0.850	1.000	6.623

A summary of weights used in each test configuration is listed in Table 8 and Figure 7 illustrates the positioning of the ballast weights.

3.8 CFD SIMULATIONS

CFD Simulations are also carried out for both stepless and three-step variants to supplement limited experimental data available[8] as well as facilitate further investigations and understanding.

[7] This reduces the effort needed to manufacture a new base plate for heave post attachment – ensuring the model is kept as similar to the original model in the stepless experiments as possible.

[8] Due to the limited time available in the towing tank, the testing of all model configurations for varying LCG position was not possible.

3.8 (a) FINE/Marine

FINE/Marine is a CFD package designed specifically for Naval Architects by NUMECA. FINE/Marine is unique in its relative ease of setup as proven by Andrun, et al. in a study of surface-piercing hydrofoils using FINE/Marine [11]. This is made possible through the C-Wizard, which automates meshing and the set-up of simulation parameters, allowing for a fast setup with excellent results. A research project found that the CFD results were within 0.7% of those from experimental tow tank tests, proving the high degree of reliability offered by FINE/Marine as a numerical tow tank [12].

3.8 (b) Turbulence Model

The k-ω (SST-Menter) model is the default turbulence model recommended by NUMECA International [13] – it combines the standard Wilcox k-ω and the standard k-ε models for a good balance between computational cost and accuracy of results [14]. An additional benefit for this study is that the k-ω SST model is especially suitable for separated flow [15].

3.8 (c) Simulation Scale

All simulations were performed at model scale. This has the benefit of being able to directly compare CFD results to experimental tests without the need to extrapolate.

3.8 (d) Simulation Test Matrix

Since stepped hulls only start becoming effective around Fn_∇=2.68, there is no value to be gained from low-speed simulations, allowing for a reduced number of tests is conducted. Table 9 summarises the simulation test matrix.

Table 9: Simulation Test Matrix

	LCG [mm]	Δ [kg]	Fn_∇ [-]
0S	408/448	6.92/4.83	2.50 – 3.00
3S	408/448	6.75/4.71	2.50 – 3.00

3.9 POST PROCESSING

Data collected from experimental test runs and numerical simulations will require post-processing for analysis. The primary result of interest is the resistance of the hull and this will be non-dimensionalised for comparison using Equation (4)

$$\text{Drag to Lift Ratio} = \frac{R_T}{W} \qquad (4)$$

3.9 (a) Processing Experimental Results

The acquisition of data in the Newcastle University Hydrodynamics Laboratory Towing Tank starts 8 metres from carriage launch and is set to record for a total of 10 seconds. At higher speeds, this is an issue as the carriage needs more than 8 metres to achieve the desired tow speed – as demonstrated by the steadily increasing velocity in the first second of Figure 8 – along with the need to to begin decelerating well before 10 seconds of data collection (shown by the steady decrease in velocity just after the 4 second mark on Figure 8) to avoid running off at the end of the tow track.

Figure 8 also demonstrates that time is required for the vessel to achieve a steady state planing condition (as evidenced from the significant changes in resistance before the 1.5 second mark) and that a relatively stable value is only achieved between the second to fourth seconds. The minute fluctuations in resistance within that range arise from a form of oscillatory factor, a result of the carriage speed control system's cyclic characteristics.

Each individual test run must thus be analysed for the criteria of speed and resistance stability. A suitable frame is identified and the recorded values of velocity, resistance, heave and pitch over time are then averaged over the defined range.

3.9 (b) Processing Simulation Results

As CFD simulations are an iterative process that seeks to guess the solution by measuring the residual value with each iteration (which quantifies the error of the guess result from the correct solution), the output must be checked for convergence to determine a useful data range. This is achieved by averaging the last 10% of results at each time step to check for convergence – defined as the data values being within 1% of the average. This is best illustrated in Figure 9, which demonstrates the changing value of trim over time-steps until convergence is achieved after approximately 2500 iterations.

Each CFD simulation is checked for convergence as defined above, and the time step at which convergence has been achieved for each individual recorded quantity (Resistance, Viscous Resistance, Pressure Resistance, Heave, Trim) is determined. The final values are then determined by averaging the last 10% of recorded data from the latest time step across all quantities for which convergence is achieved.

4. RESULTS & DISCUSSION

In this section, the results obtained from both experimental and numerical testing are analysed and the significance of them discussed.

4.1 ACCURACY OF CFD RESULTS

The first order of business is to verify the accuracy of the results obtained from CFD simulations. To this end, existing results from Bowles' original model tests are compared with results from the CFD simulations for a stepless model at various conditions. As can be seen from Table 10, the CFD readings show an excellent correlation with the experimental readings, with an average error of less than 2% – with the exception of TC4 – proving the validity of the CFD model in closely simulating actual physical conditions.

Table 10: Stepless CFD vs Experimental Results

	Fn_∇ [-]	R_T [N] CFD	R_T [N] Experimental	Difference
0S-TC1	2.50	12.310	12.147	1.342%
	2.75	12.292	12.050	2.010%
0S-TC2	2.50	13.659	14.046	-2.755%
	2.75	13.676	13.593	0.613%
0S-TC3	2.75	9.105	8.891	2.410%
	3.00	9.047	8.952	1.058%
0S-TC4	2.75	8.005	9.069	-11.728%
	3.00	8.232	9.159	-10.122%

The source of the relatively higher margin of error between CFD and experimental results for TC4 arises from the instability of the vessel at that particular set up. Analysing Figure 10, it can be seen that the model exhibits a porpoising behaviour which is characteristic of light planing hulls with a longitudinal centre of gravity located too far back. No convergence of results is achieved. This in turn significantly affects the resistance value measured from the simulations – since the value is taken by averaging the last 10% of recorded data – causing an inaccurate recording for the simulations.

Unfortunately, raw data from the towing tank experiments for the original stepless planing hull are no longer available, so it is not apparent if the physical model exhibited the same porpoising phenomenon; but it is more than likely that this lack of a steady planing state is the root cause of the rather significant difference in experimental and simulation results. The findings from this validation test verify the simulation parameters in use and allow the CFD results for the stepped models to be used with confidence.

4.2 EXPERIMENTAL FINDINGS

In this study, two configurations of the model are tested at varying masses – i.e. loading conditions – for comparison with the basis stepless model. Three sets of experiments are conducted for each test configuration, with results averaged across the sets for an accurate result. Due to the limited time scope available for experimental testing, a variation in LCG is not achievable. Simulations (presented later) are performed to achieve a complete set of data. Results from experimental findings will now be presented in the next two sub-sections.

4.2 (a) Test Configuration 1 (Experimental)

Referring to Figure 11, it is immediately apparent that for the heavier loading scenario of TC1, the three-step model exhibits a larger Resistance-to-Weight ratio than the stepless model at Volumetric Froude Numbers above about 1.3. Although the three-step model does exhibit a lower R_T/W at speeds below Fn_∇=1.3, this is of little impact since planing vessels operate at high speeds and this decrease in resistance at lower speeds is of little significance. It is then worrying that at higher speeds, the three-step model exhibits increasingly larger resistance as compared to the stepless model.

At Fn_∇=2.75, the three-step model has an increased R_T/W of 16.88% as compared to the basis stepless model. This could potentially be due to the absence of flow separation. There are a number of reasons for this: a lack of time to achieve separation of flow at steps (since separation requires time and the tow tank could be too short for this to be attained) a failure to ventilate steps due to the heavy loading of the vessel corresponding to a large draft. With there being no separation, a stepped model would result in a larger Wetted Surface Area – arising from the increase in hull area due to the step itself – and thus increased frictional resistance. The removal of hull volume also causes a reduction in volume aft of the vessel, resulting in an increased trim which would then increase pressure resistance. It is then unsurprising that the possible failure to attain flow separation in this experimental configuration has resulted in the drastic increase in resistance of the three-step model over a basis stepless model.

It is also interesting to note the kink in the three-step plot at Fn_∇=2.25. At this speed, the three-step model has a 25.48% increase in RT/W over the stepless model, after which the resistance increase drops significantly to 14.27% at Fn_∇=2.50. This could potentially be due to the fact that the inception of flow separation at steps begins to come into effect at approximately Fn_∇=2.50. Underwater footage, while slightly blurry, illustrates this with a white streak seen around the first step position in the Fn_∇=2.50 frame grab (Figure 12).

Since this streak is not visible in the Fn_∇=2.25 frame grab, it is appropriate to conclude with reasonable confidence that this indeed is an illustration of flow separation beginning to occur at the step. The observance of flow separation then debunks the theory of there being no flow separation at all, although it is possible that the flow separation is insufficient for the steps to become effective. From the frame grabs, it is impossible to determine the extent of reduction in wetted surface area arising from the separated flow. It is quite probable that flow separation is only occurring at the first step since separation of flow occurs sequentially down the steps – again returning to the point of the towing tank not being long enough for the model to fully achieve a steady planing state with separation attained at all steps. This will be further discussed later together with the simulation results.

4.2 (b) Test Configuration 3 (Experimental)

Compared to those of TC1, the results illustrated in Figure 13 shows show a considerably smaller increase in Resistance-to-Weight ratio. This is due to the lighter vessel being able to attain a steady planing state much easier than the heavier model (along with achieving complete planing over a shorter distance). Examining the vessel Centre of Gravity rise provides some insight into this. From Figure 14, it is clear that the three-step variant of the lighter TC3 setup shows considerably larger CG rise (~66% higher than stepless TC3 at $Fn\nabla$=2.75) than the three-step variant of the TC1 configuration (~13% higher than stepless TC1 at $Fn\nabla$=2.75). The additional lift generated by flow reattachment to the hull – aft of the step – aids the lighter vessel more in elevating the hull out of the water for planing.

Another possibility could also be due to the lighter draft facilitating increased step ventilation and thus flow separation at higher speeds. However, without clear underwater footage to examine the hull lines along with the nature of flow around and under the hull, it is impossible to compare the wetted surface areas between the two model configurations to confirm this.

The kink in the resistance-time plot is more pronounced for TC3, verifying the earlier proposed theory in Section 4.2 (a) that flow separation at steps does indeed begin somewhere in the range of Fn_∇=2.25-2.50.

At Fn_∇=3.00, the R_T/W of the three-step variant dips below the basis stepless hull by 0.3%, in line with earlier studies that a stepped hull would become beneficial over a stepless hull between $Fn\nabla$=2.68 and $Fn\nabla$=4.00. A three-step hull can thus at least offer the same performance improvements as a single-step hull.

4.3 CFD SIMULATION FINDINGS

As validated in Section 4.1, the CFD model configured offers excellent and accurate results with negligible differences from the experimental results of a stepless model. This can thus be used to overcome the difficulties faced in experimental testing of the three-step model, such as insufficient length of towing tank to attain a steady planing state and complete flow separation at steps. An added benefit of CFD simulation is the possibility to obtain the breakdown of resistance between pressure and viscous components directly. The results from simulations conducted will now be presented in this section.

Table 11: CFD Simulation Results

	Fn_∇ [-]	R_T/W [-]	R_P/W [-]	R_V/W [-]	Trim [deg]	WSA [m^2]
0S						
TC1	2.50	0.181	0.086	0.096	3.982	0.159
	2.75	0.181	0.085	0.096	3.965	0.159
TC2	2.75	0.201	0.123	0.078	6.646	0.123
	3.00	0.201	0.123	0.078	6.651	0.122
TC3	2.50	0.191	0.067	0.124	2.467	0.145
	2.75	0.192	0.067	0.126	2.379	0.148
TC4	2.75	0.174	0.077	0.096	4.081	0.092
	3.00	0.174	0.082	0.091	4.276	0.097
3S						
	Fn_∇ [-]	R_T/W [-]	R_P/W [-]	R_V/W [-]	Trim [deg]	WSA [m^2]
TC1	2.50	0.238	0.157	0.080	5.595	0.146
	2.75	0.238	0.157	0.080	5.588	0.146
TC2	2.75	0.196	0.163	0.033	8.180	0.059
	3.00	0.198	0.165	0.033	8.194	0.059
TC3	2.50	0.173	0.113	0.061	5.087	0.061
	2.75	0.171	0.111	0.061	5.073	0.062
TC4	2.75	0.173	0.125	0.048	6.121	0.049
	3.00	0.175	0.127	0.048	6.136	0.049

Examining Table 11 shows that a three-step variant offers improvements in the Resistance-to-Weight ratio in all cases but TC1 (the heaviest configuration with the LCG pushed further forward) – where the large increase in trim causing a significant increase in pressure resistance outweighs the reduced viscous resistance due to the reduced wetted surface area. Recreating a visualisation of the 3S-TC1 simulations shows that flow separation is achieved only momentarily as the vessel builds up to its steady planing state, which accounts for the increased R_T/W. This is due to the LCG being located too far forwards, resulting in the steps becoming too far submerged to attain ventilation as the vessel trims about its LCG.

In all other setups, an increase in trim is seen, congruent with the theory of lost buoyancy aft and resulting in an increase in pressure resistance. However, flow separation achieved at the steps (as illustrated in Figure 15) has allowed for a significant reduction in WSA, thus giving rise to a more impactful decrease in viscous resistance. The overall effect is a reduction in R_T/W of up to 11%. This further supports the trend determined in experimental testing. Again, recreating a visualisation of the three-step simulations offers further insights into the vessel performance. In all cases, a steady state of flow separation is only achieved after at least approximately 2.5 to 3 seconds of physical time. The renderings also demonstrate that flow separation does indeed occur sequentially with flow separating first from the forward most step before progressing to subsequent steps as evident from Figure 16.

Cross-examining the experimental data for the high-speed runs show that the model only has about 2.6 seconds of data acquisition (after attaining a planing state) before it hits the stop switch and begins decelerating. There is insufficient time for complete flow separation to be attained in the physical tow tank tests, accounting for the higher than expected resistance results collected experimentally. As presented in Section 4.2 (a) with reference to Figure 12, the physical models only manage to attain fully developed flow separation at the first step within the testing frame, which leads to the conclusion that the three-step CFD simulation results – based on the same model as the stepless simulations which showed excellent correlation to experimental results – are more accurate than the results obtained via towing tank model tests.

It is worth noting that the design of the three-step hull still requires optimisation to achieve the least possible wetted surface. In this study, the initial assumption used was that flow separation would remain roughly equal at each step with equal step heights and lengths. As can be seen from Figure 15, this is not the case and the flow reattachment length becomes shorter with each subsequent step. It is clear then that an optimal three-step design would require a variation of step height and an unequal distribution of step lengths to seek the best resistance performance.

Moreover, the three-step design in this study was created for a maximum operational speed of Fn_∇=2.75 (to suit the Newcastle University towing tank's max carriage speed). It is clear from Figure 16 that flow reattachment occurs right at the aft limit of the first step. Running the vessel at speeds beyond Fn_∇=2.75 would only result in a detrimental effect of degrading performance, due largely to an undesirable effect whereby flow separation from the first step only re-joins after the end of the first step. Flow separation at subsequent steps is compromised (potentially resulting in a non-optimal WSA) and the total lifting force arising from flow reattachment is reduced (increasing the trim). This is evident from both the 3S-TC2 and 3S-TC4 cases, whereby the vessel trim at Fn_∇=3.00 is higher than at Fn_∇=2.75, leading to an increased pressure resistance (as presented in Table 11). A three-step configuration designed for higher speeds may well prove to offer better results throughout the planing regime.

It is also worth noting that for stepless hulls in general, for the heavier loading configurations, positioning the LCG further towards the aft results in higher R_T/W; whereas for lighter configurations, positioning the LCG towards the aft offers a lower R_T/W. There is however no striking relationship between the LCG position and resistance performance for the three-step model.

5. CONCLUSIONS

This study has examined the resistance characteristics of a three-step hull through a dual-pronged experimental and numerical approach. Initial stepless numerical simulations established a model to be used that produced an excellent agreement with the baseline stepless experimental results, proving the reliability of the model for subsequent multi-step simulations. This facilitated overcoming physical and time constraints of experimental testing to provide a more complete picture.

While experimental testing of a three-step model showed poor resistance characteristics (i.e. a higher R_T/W as compared to the basis stepless model), careful examination of underwater footage and CFD simulations exposed the unreliability of experimental data. It was shown that flow separation takes time to achieve; in this case, the Newcastle University Hydrodynamics Laboratory Towing Tank was too short for the testing of high-speed craft and the three-step hull model ultimately failed to attain complete flow separation in the short time span it takes the carriage to travel the tank's length. Simulations then show that for configurations where flow separation can be successfully achieved, the three-step model indeed offers improved resistance performance over a basis stepless model. This investigation also concluded that an optimised three-step design with varying step height and length distribution would quite likely produce better results.

It was also proved that Svahn's method for Stepped Hull Performance Prediction cannot simply be extended to a multitude of steps, even though it does offer a good basis for initial design estimation – demonstrated by the close alignment of flow reattachment length observed at the first step with predictions using Svahn's method.

5.1 RECOMMENDATIONS

Following this research, there are still numerous uncertainties that can be better clarified in the search for an optimum stepped-hull design. Efforts to vary multi-step parameters (i.e. height, position, etc.) should be made to determine if a well-designed three-step hull can match or better the resistance performance of a single-step hull. Furthermore, there is scope to extend this study by examining if the multi-step hulls can bring about improvements in seakeeping performance through a distribution of lifting forces over the steps. Experimental testing in a longer towing tank can also aid in further validating the simulation results. As this study specifically designed a multi-step hull with certain assumptions including a maximum operating speed of Fn_∇=2.75, there is also potential to conduct a more extensive set of tests at higher speeds to properly determine the influence of different multi-step parameters on vessel performance. Investigating all the above will aid in establishing if a true limit to the number of steps (before benefits begin diminishing) exists.

6. ACKNOWLEDGEMENTS

I would like to thank my undergraduate dissertation supervisor, Dr David Trodden, for his invaluable support throughout the course of this project, for always taking time out of his busy schedule to accommodate flexible meetings and discussions, and for his trust in me to tackle the scope of this project. Many thanks also go to the team of technicians in the Hydrodynamics Laboratory, whom kindly sat down with us early in the academic year to discuss experimental options for this project and helped in various aspects of the experimental conduct including manufacturing of the stepped model. Last but not least, I must express my gratitude towards Mr Jeffery Bowles, for graciously sending me his report from his original planing hull tests.

7. REFERENCES

1. CLEMENT, E. P. & KOELBEL, J. G., 'Optimized Designs for Stepped Planing Monohulls and Catamarans', *American Society of Naval Engineers, pp. PC35-PC44*, 1992.

2. GARLAND, W. R., 'Stepped Planing Hull Investigation', *Transactions - Society of Naval Architects and Marine Engineers pg 448-458*, 2011.

3. LONI, A., GHADIMI, P., NOWRUZI, H. & DASHTIMANESH, A., 'Developing a computer program for mathematical investigation of stepped planing hull characteristics', *International Journal of Physical Research, 1(2), pp. 34-47*, 2013.

4. LEE, E., PAVKOV, M. & MCCUE-WEIL, L., 'The Systematic Variation of Step Configuration and Displacement for a Double-step Planing Craft', *Journal of Ship Production and Design, May, 30(2), pp. 89-97*, 2014.

5. SAVITSKY, D., 'Hydrodynamic Design of Planing Hulls', *Marine Technology, October, 1(1), pp. 71-95*, 1964.

6. SAVITSKY, D. & MORABITO, M., 'Surface Wave Contours Associated with the Forebody Wake of Stepped Planing Hulls', *Marine Technology, 47(1), pp. 1-16*, 2010.

7. SVAHN, D., 'Performance Prediction of Hulls with Transverse Steps', *Stockholm: KTH Royal Institute of Technology*, 2009.

8. CLEMENT, E. P., 'Analyzing the Stepless Planing Boat', *Washington, DC: David Taylor Model Basin*, 1956.

9. BOWLES, J. B., 'An Investigation into the Effects of Waterjet Tunnels on the Resistance, Trim, and Dynamic Lift of a Planing Monohull', *Newcastle upon Tyne: The University of Newcastle upon Tyne*, 2001.

10. 'Hydrodynamics Laboratory', *Newcastle University*, 2019.

11. ANDRUN, M., ŠARIĆ, B., BASIC, J. & BLAGOJEVIĆ, B., 'CFD Analysis of Surface-Piercing Hydrofoil Ventilation Inception', *XXII Theory and Practice of Shipbuilding*, 2016.

12. MALLOL, B., 'Marintek chooses FINE™/Marine for planing hull resistance curve prediction', *NUMECA*, 2016.

13. 'Best Practice for Turbulence Modeling', *NUMECA International*, 2018.

14. 'Menter Two-Equation model (SST)', *NUMECA International*, 2018.

15. MENTER, F. R., 'Improved Two-Equation k-ω Turbulence Models for Aerodynamic Flows', *National Aeronautics and Space Administration*, 1992.

8. AUTHORS BIOGRAPHY

Junwei Fan graduated with a BEng from Newcastle University in 2019 as Best Graduating Student in Naval Architecture. He is currently a post-graduate student pursuing a M.S. Naval Architecture & Marine Engineering at the University of Michigan, and is continuing research on stepped planing hulls for his Master's thesis.

Dr David Trodden holds the current position of Lecturer in Marine Technology at Newcastle University. He is responsible for the teaching of modules such as Ship Design, and Advanced Marine Vehicles. His previous research experience and interests as a ship hydrodynamicist include ship resistance and propulsion, advanced model measurement techniques, and ship simulations.

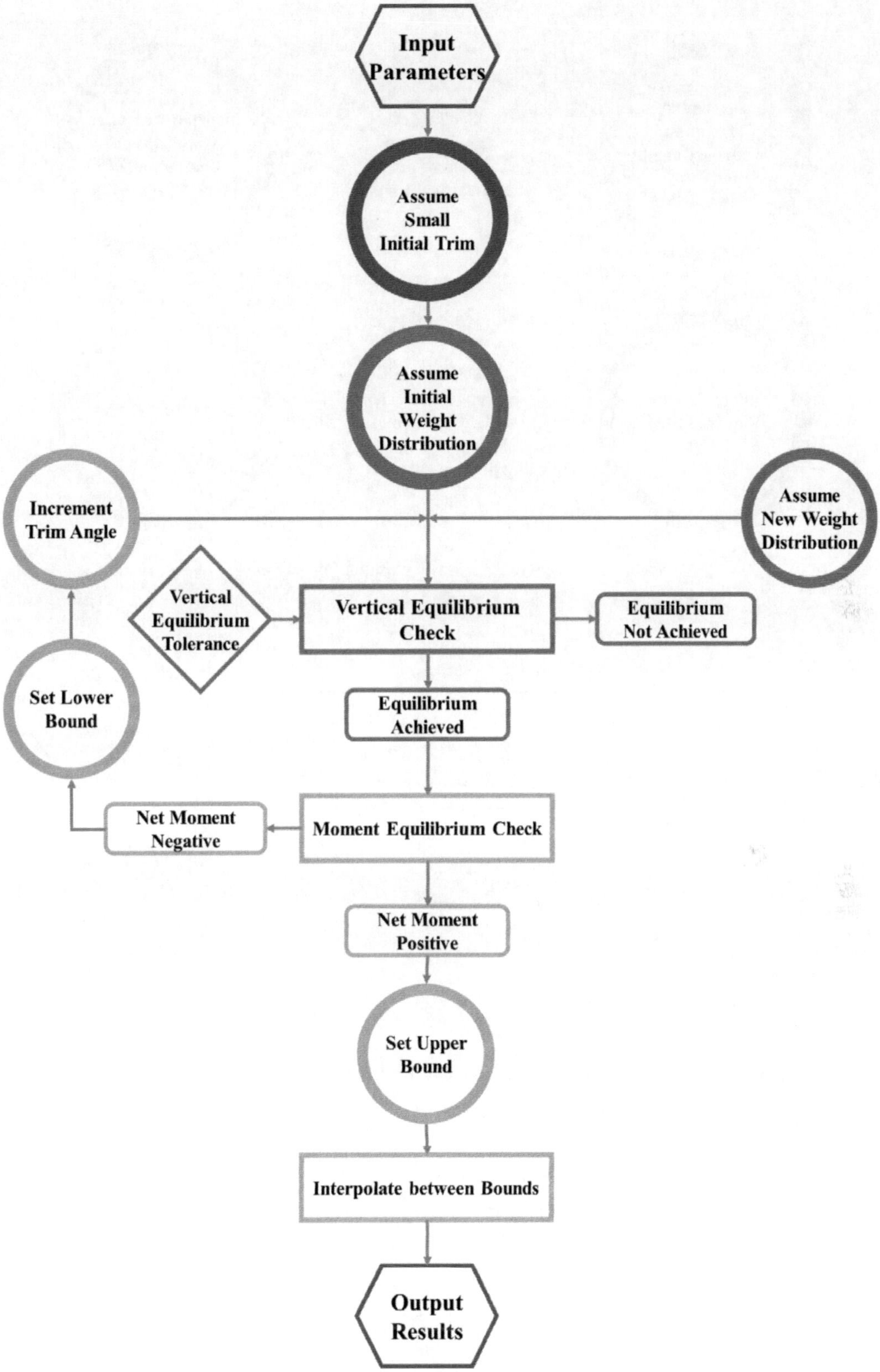

Figure 1: Illustration of Program Flow implementing Svahn's Method

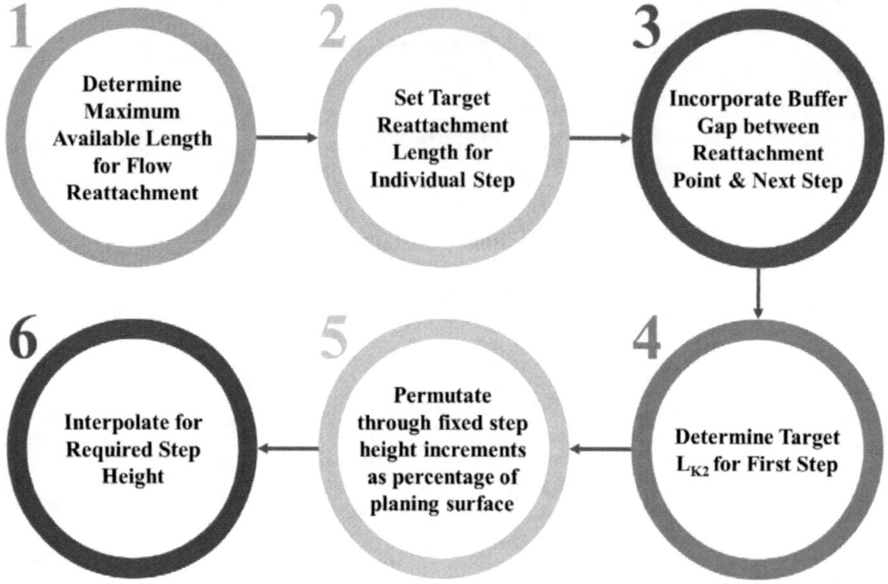

Figure 2: Procedure for Flow Separation Investigation

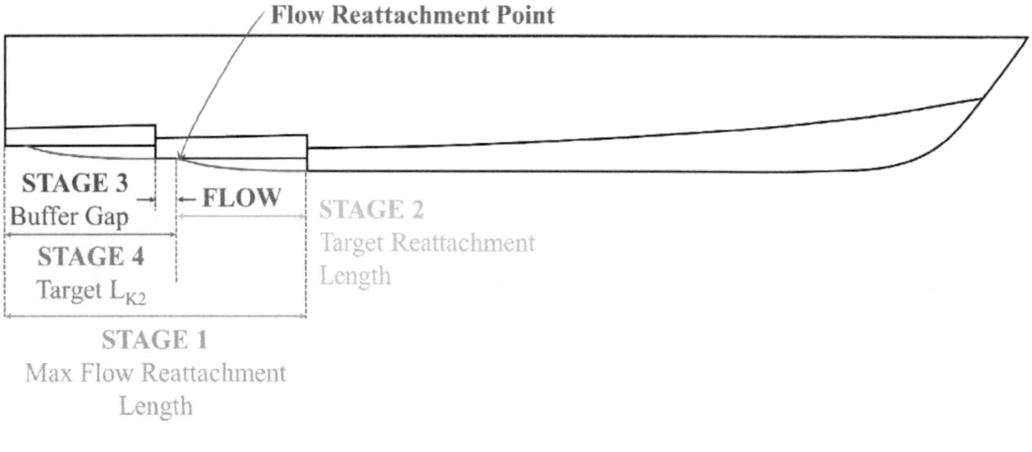

Figure 3: Illustration of Resulting Stage Parameters from Flow Separation Investigation

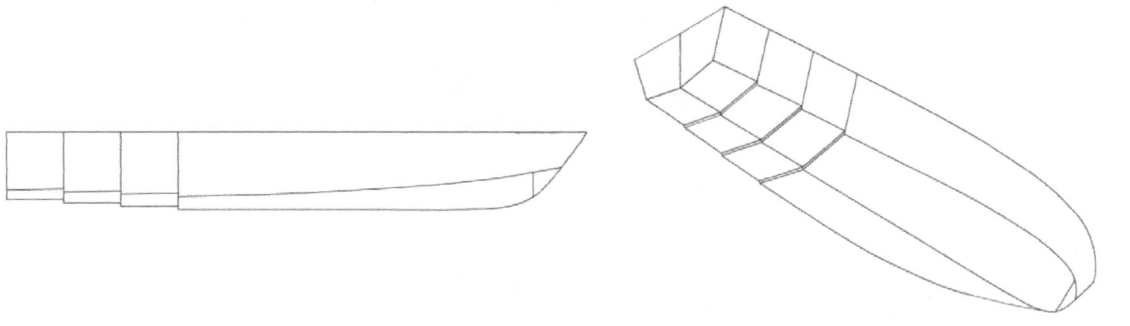

Figure 4: Three-step Hull Variant Illustration

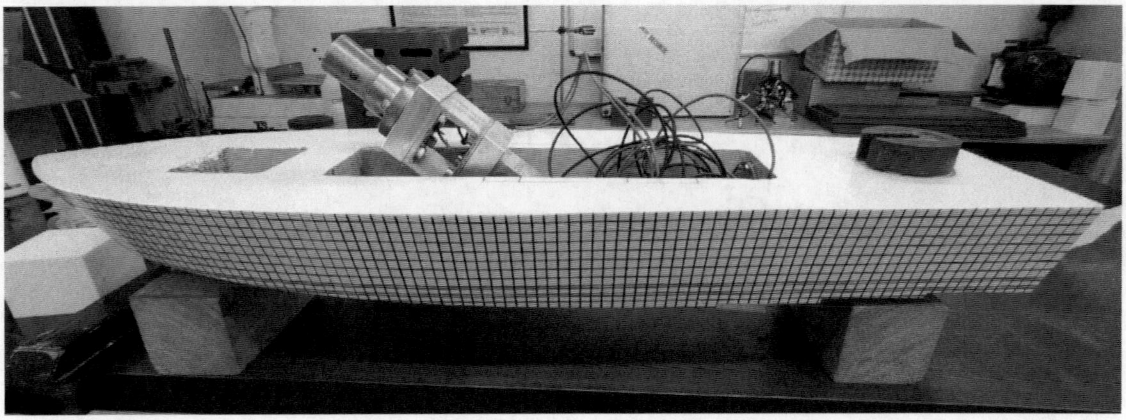

Figure 5: Modified 4667-1 with Three Steps

Figure 6: Restriction of Counter-Weight Movement

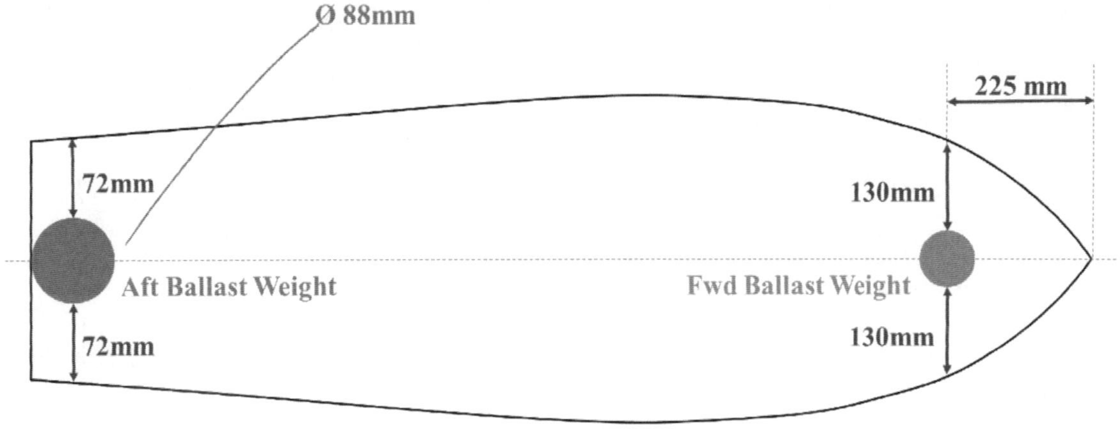

Figure 7: Ballast Weight Positioning (Top View)

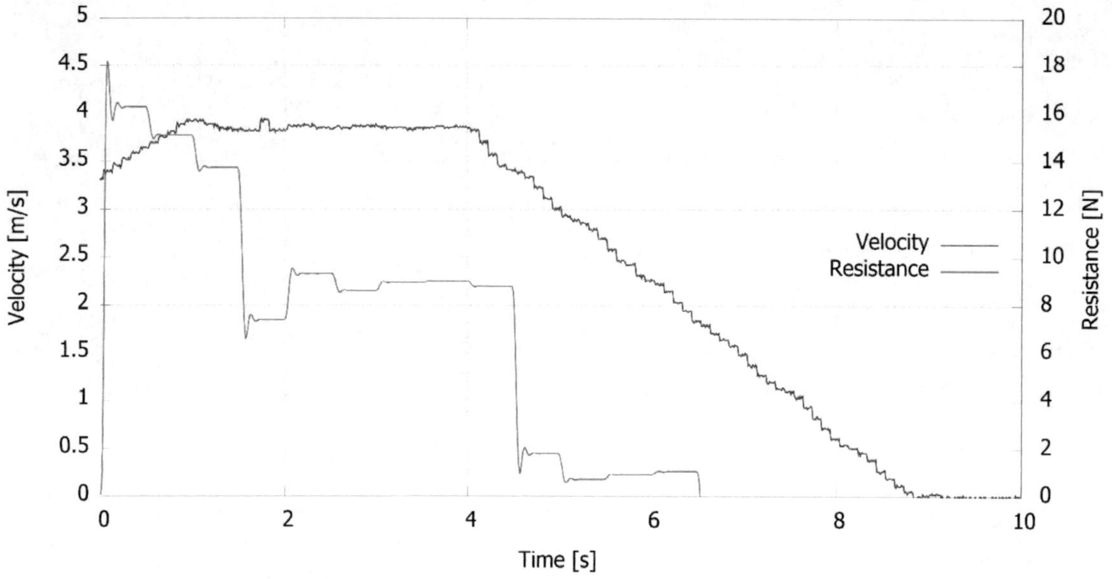

Figure 8: Time-Series Plot of Velocity & Resistance for Experimental 3S-TC3 @ 3.847 ms^{-1} (Fn$_\nabla$ = 3.00)

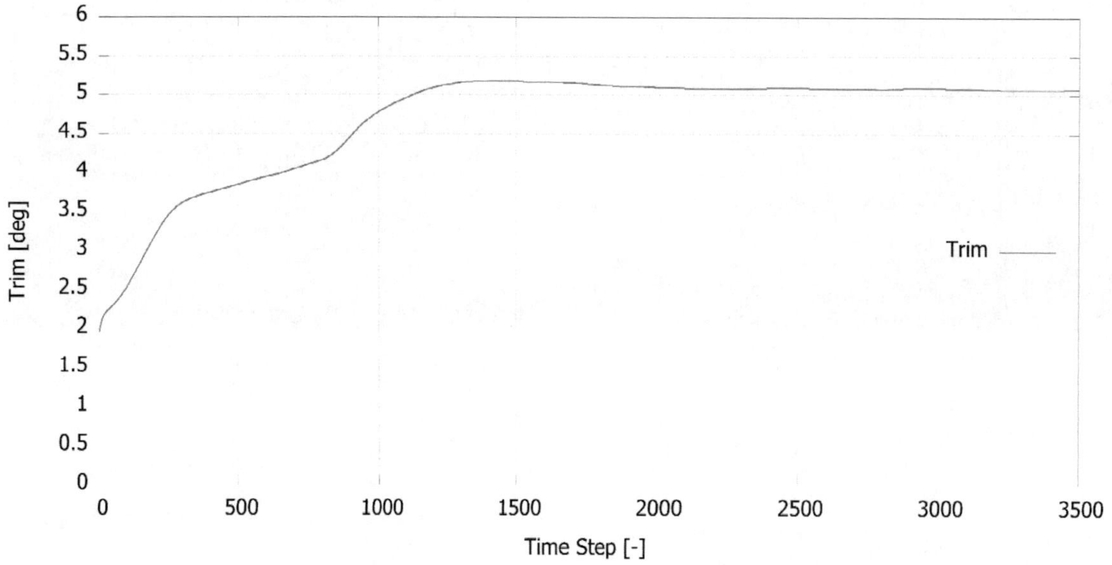

Figure 9: Time-Series Plot of Trim for Simulation 3S-TC3 @ 3.848 ms^{-1} (Fn$_\nabla$ = 3.00)

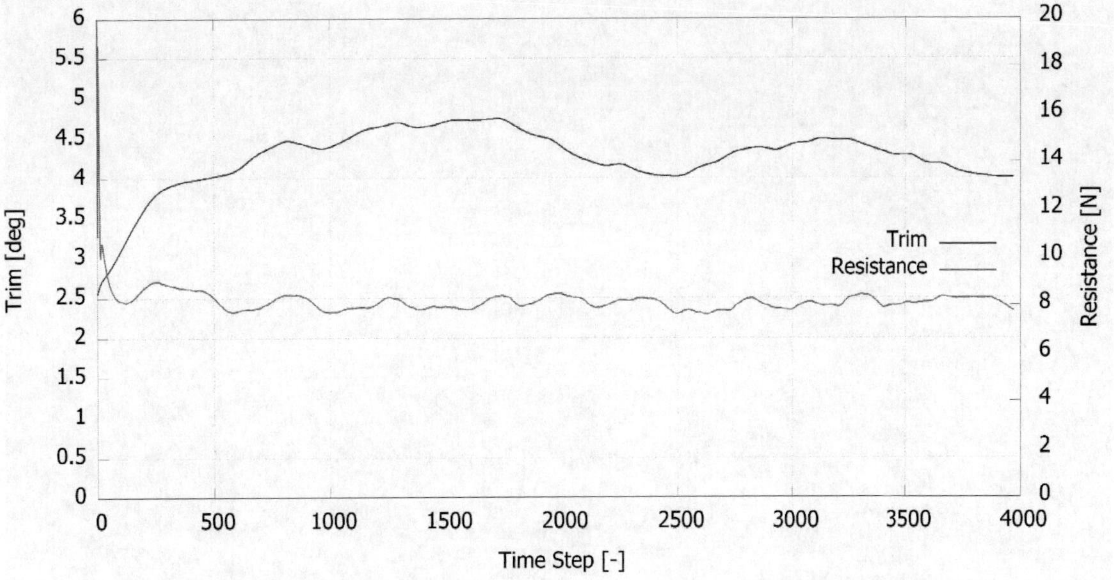

Figure 10: Time-Series Plot of Resistance & Trim for 0S-TC3 @ 3.542 ms^{-1} (Fn$_\nabla$= 2.75)

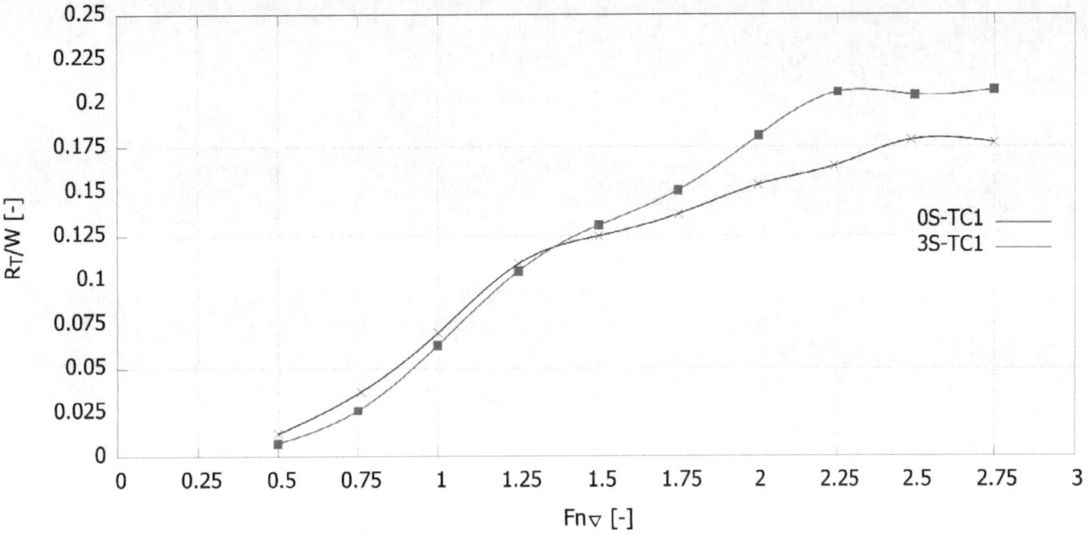

Figure 11: Comparison of Experimental Resistance to Volumetric Froude Number between 0S and 3S for TC1

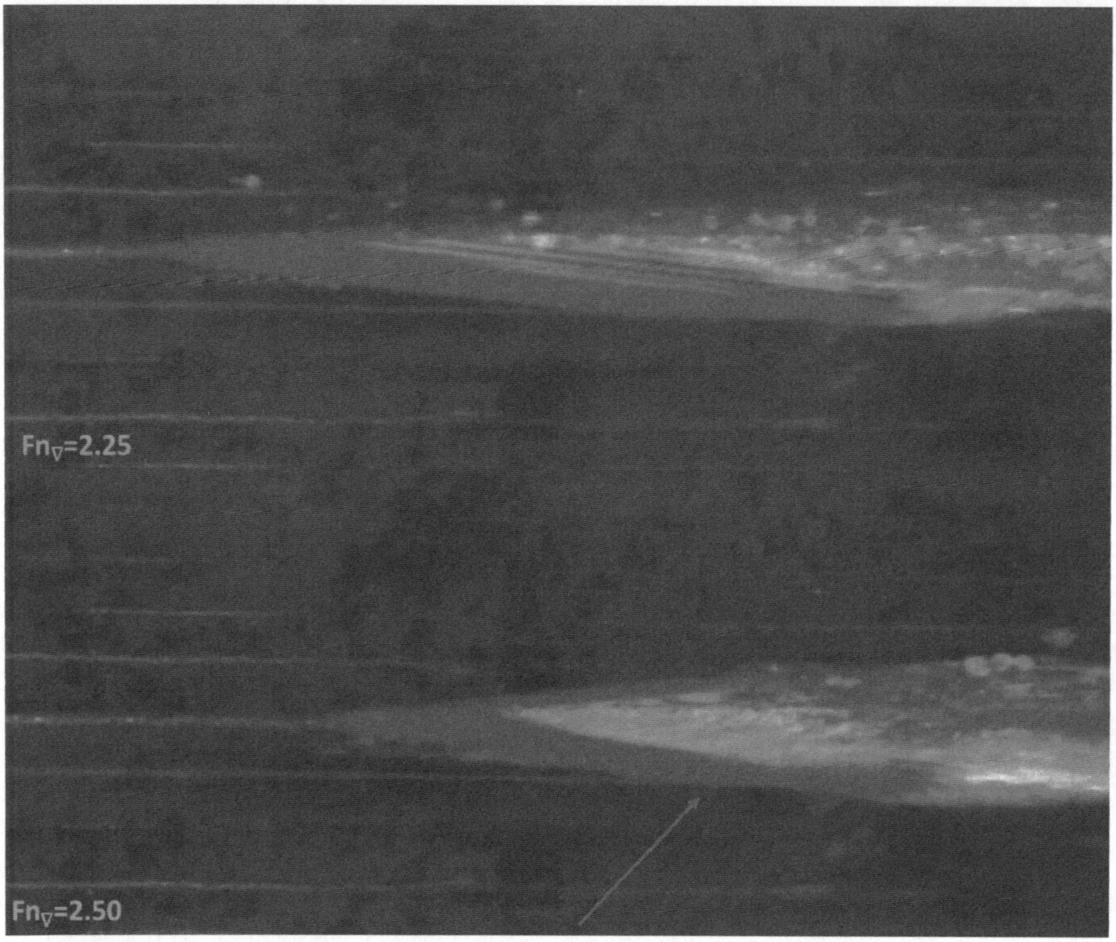

Figure 12: Underwater Camera Frame Grab of 3S-TC1 @ Fn∇ = 2.25 & Fn∇ = 2.50

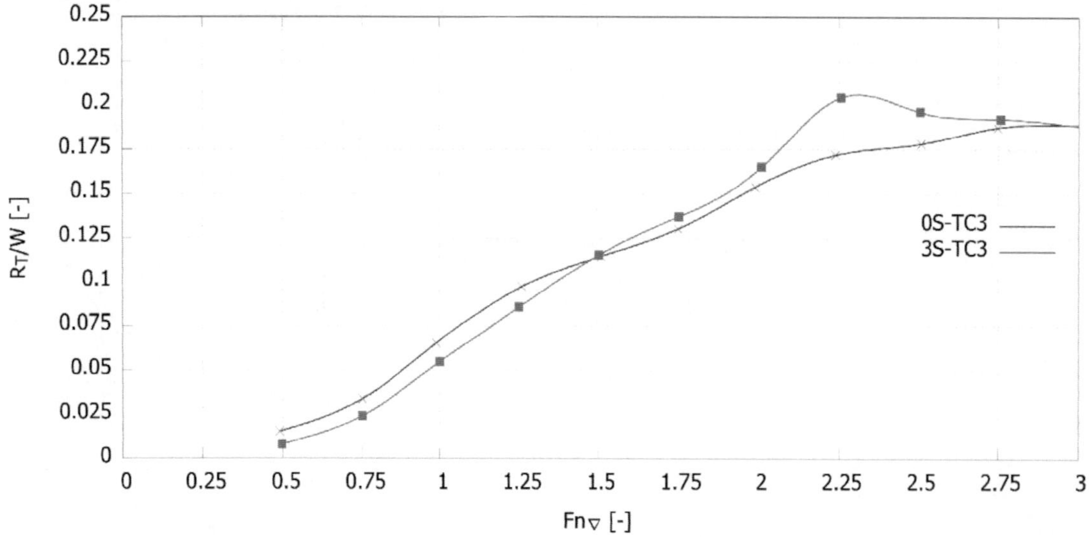

Figure 13: Comparison of Experimental Resistance to Volumetric Froude Number between 0S and 3S for TC3

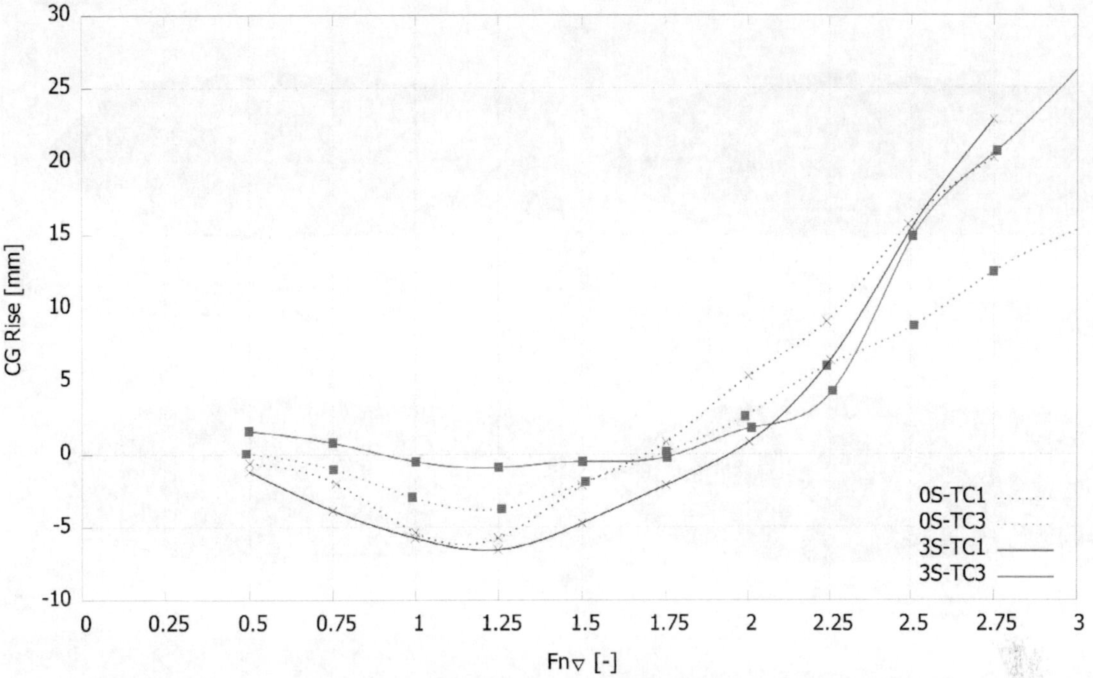

Figure 14: Comparison of Experimental Heave to Volumetric Froude Number between 0S and 3S for TC1 & TC3

Figure 15: Free Surface Illustration for 3S-TC3 @ 3.527 ms^{-1} ($Fn_\nabla = 2.75$)

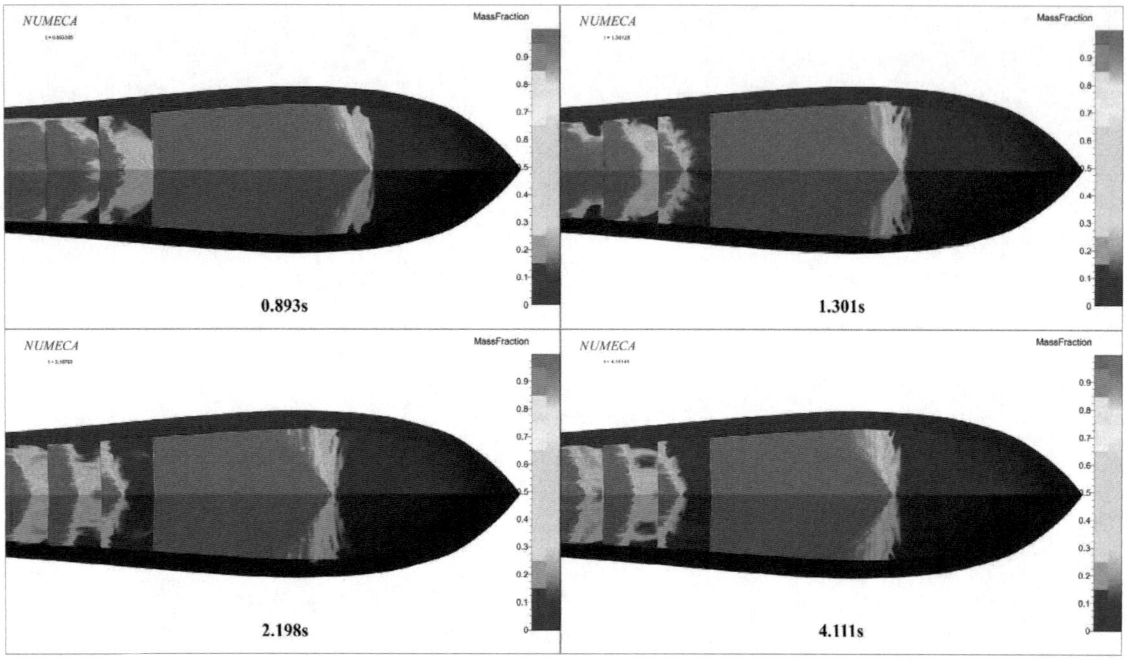

Figure 16: Progress of Flow Separation for 3S-TC2 @ 3.744 ms^{-1} ($Fn_\nabla = 2.75$)

RESISTANCE TESTS OF A SERIES OF CHINE INTERCEPTORS ON A PLANING CRAFT

M. Lakatos, **T. Sahk** **H. Andreasson** and **K. Tabri**, Tallinn University of Technology, Estonia

SUMMARY

Information on chine interceptors, as well as the whole topic of spray sheet deflection, is scarce and to the most extent dated. Therefore, a series of model tests have been conducted with a 19 m craft configured with chine interceptor arrangements (operating in a range of Froude numbers Fr = 0.77…1.35). The series is comprised of two benchmarks and two modifications for each of the benchmarks, altogether 6 models. Each model was tested at four displacements and speeds. The first benchmark, a bare hull, had the following two chine modifications: a 2x2mm rectangular rail attached along the outer edge of the chine and a 2x15mm chine inclination that covered the chine, resulting in a new chine surface angled down. The second benchmark, a hull equipped with spray rails with a rectangular 5x5mm profile, was tested with a rectangular 2x2mm and a triangular 5x5mm profile attached at the outer edge of the chine. Chine interceptors on reduced the total resistance of a planing craft by more than 3% both on a bare hull and hull equipped with spray rails.

NOMENCLATURE

L_{OA}	Length overall (m)
L_{PP}	Length between perpendiculars (m)
L_{WL}	Length waterline (m)
B_{OA}	Beam overall (m)
B_{WL}	Beam waterline (m)
T	Draught (m)
Δ_m	Displacement mass (kg)
∇	Displacement volume (m^3)
L_{CG}	Longitudinal Centre of Gravity (m)
KG	Vertical Centre of Gravity (m)
Fr	Froude number (1)
Fr_∇	Froude displacement number (1)
R_{TM}	Total resistance in model scale (N)
τ	Dynamic trim (degrees)
$z_{V,CG}/\ \nabla^{1/3}$	Non dimensional dynamic sinkage (1)
BH	Bare hull
SR	Hull with spray rails
CSR	Chine spray rail
CSI	Chine spray interceptor
C.INC	Chine inclination

1. INTRODUCTION

Whisker spray is a phenomenon that occurs on high-speed marine vehicles (semi-displacement craft, planing craft and hydroplanes) when they operate at high speeds. It is created from the stagnation line (a line that separates the flow going under the hull into the pressure area from the flow going into the spray area) and it forms a thin sheet of fluid along the bottom surface above the stagnation line, which under low-speed operation would have been dry. This spray sheet of water causes frictional resistance along its streamline over the hull surface. The resistance of whisker spray has been shown to account for 10% - 25% of total resistance (at high speed, when Froude displacement number $Fr_\nabla > 4$) by inference from a comparison of model test data with and without spray rails installed on the hull bottom [1]–[4].

To reduce the whisker spray resistance, two to four longitudinal stripes (also called rails, lifting strikes or deflectors) are usually installed on the hull. The main purpose of these appendages is to detach the spray from the hull surface and deflect it towards the sides or slightly down and aftward, thereby reducing the frictional resistance by up to 18%. Deflection of the spray also induces a lifting force that affects the running trim of the craft. Depending on the position and shape of the spray rail as well as the orientation of its deflecting surface, it is possible to generate different amounts of lift resulting from the whisker spray deflection [5]. Since the running trim has a more significant effect on the total resistance than a whisker spray, a spray rail will reduce the resistance due to its lift producing qualities rather than due to its spray detachment abilities.

An analytical model [6] that predicts the viscous drag in the spray area as a function of deadrise angle, trim angle and speed has been added to the Savitsky method [7], which previously only included the viscous and pressure drag components in the bottom area aft of the stagnation line. In addition to whiskers spray, there is also a second spray pattern called main spray [8], observed in planing crafts. It is a discharge of water in the form of a cone with its apex located near the intersection of the stagnation line and the chine. The outboard trajectory of the main spray is significantly elevated compared to that of the whisker spray. Due to the large volume of water and the high trajectory angle of the main spray, it can impact the wings and propellers of seaplanes causing structural damage. It can also impact the tunnel between the hulls of a planing catamaran and thereby significantly increase in resistance. Savitsky and Morabito [9] developed and experimentally verified an analytical model that predicts the geometry of the main spray as a function of deadrise angle, trim angle and speed.
A few studies featuring spray rails on planing hulls have been published within the last fifteen years. However, in most of those studies, the focus is either on the validation

of numerical models of planing hulls in calm water [10] and waves [11] or the evaluation of new hull form concepts [12]–[14].

A recent study [15] showed that chine modifications can reduce the total resistance of a planinig hull by up to 3.5%. However, the hull featured in that study had no spray rails, which begged the question whether the effect of the chine modifications would be the same on a hull with spray rails. Therefore, this study investigates the influence of the chine spray rails and interceptors on the total resistance trim and sinkage of a planing craft already equipped with spray rails.

2. EXPERIMENTAL TESTS

2.1 DESCRIPTION OF TESTED MODELS

A test case craft was designed and manufactured by the Small Craft Competence Centre (SCC) in Kuressaare, Estonia, to compare conventional and novel spray rail configurations. The main particulars (Table 1) of the hull design are common for a high-speed patrol, search and rescue (SAR) craft or a larger pleasure craft. The craft was designed to operate at a Fr = 1.354 and Fr_∇ = 3.11, equivalent to the displacement of 40 t and speed of 35 knots in full scale.

Table 1: Main dimensions of the naked hull.

Parameter		
Model Scale	λ	10
Length overall	L_{OA}	1.921 m
Length between perpendiculars	L_{PP}	1.800 m
Length waterline	L_{WL}	1.703 m
Beam overall	B_{OA}	0.581 m
Beam waterline	B_{WL}	0.581 m
Draught	T	0.108 m
Displacement mass	Δ_m	40 kg
Displacement volume	∇	0.040 m3
Longitudinal Centre of Gravity	L_{CG}	0.669 m
Vertical Centre of Gravity	KG	0.200 m

The series is comprised of two benchmarks and two modifications for each of the benchmarks, altogether 6 models. The first benchmark, a bare hull (BH) (Fig. 1a), had the following two chine modifications: a 2x2mm rectangular interceptor (Fig. 1b) attached along the outer edge of the chine (BH-CSI) and a 2x15mm chine inclination (Fig. 1c) that covered the chine, resulting in a new chine surface angled down (BH-C.INC). The second benchmark (SR), a hull equipped with spray rails (Fig. 2a) with a rectangular 5x5mm profile (Fig. 2b), was tested with a rectangular 2x2mm (Fig. 2c) chine spray interceptor (SR-CSI) and a triangular 5x5mm (Fig. 2d) chine spray rail attached at the outer edge of the chine (SR-CSR). Each configuration was tested at four speeds corresponding to full scale (20, 25, 30 and 35 knots) at the model scale of 1:10 and four loading conditions (30 kg, 35 kg, 40 kg and 45 kg displacement).

Table 2: Test matrix of the spray rail series.

Fr [-]	Δ_m **[kg]**	**Hull version**	**Description**
0.769 0.961 1.153 1.345	30 35 40 45	BH	Bare hull
		BH-CSI	Bare hull with a rectangular 2x2 mm chine spray interceptor
		BH-C.INC	Bare hull with a triangular 2x15 mm chine inclination
		SR	Hull with rectangular 5x5mm spray rails
		SR-CSI	Hull with 5x5mm rectangular spray rails and a 2x2mm rectangular chine spray interceptor
		SR-CSR	Hull with 5x5mm rectangular spray rails and a 5x5mm triangular chine spray rail

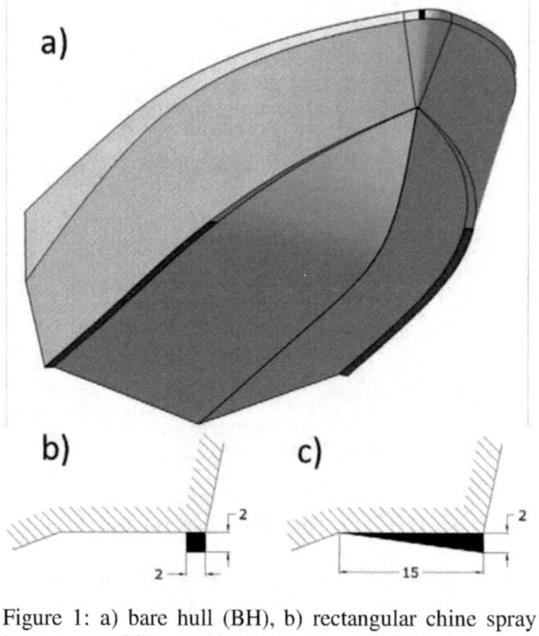

Figure 1: a) bare hull (BH), b) rectangular chine spray interceptor (CSI), c) chine inclination (C.INC).

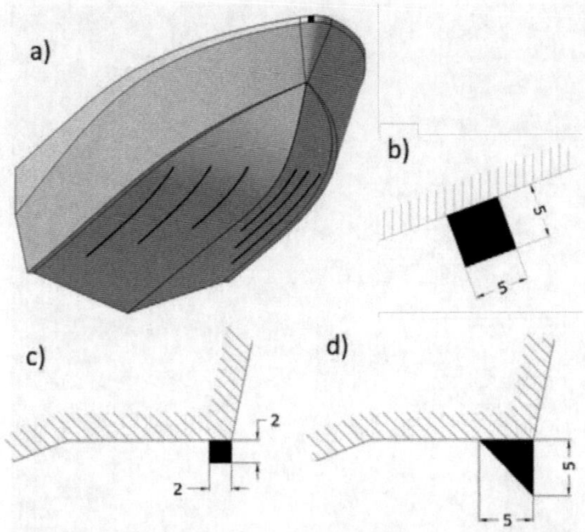

Figure 2: (a) hull with spray rails (SR), b) rectangular 5x5mm spray rail, c) rectangular 2x2mm chine spray interceptor (SR-CSI), d) triangular 5x5mm chine spray rail (SR-CSR).

2.2 LABORATORY INSTRUMENTATION

The calm water towing tests were conducted in the towing tank of SCC following the ITTC guidelines 7.5-02-02-01 [16]. The main dimensions of the basin are length 60m, breadth 5m and depth 3m. The models were tested in semi-captive condition (Fig. 3), i.e. the models were free to sink and pitch. During the tests, resistance and draught changes were measured.

The total resistance of the models was measured at 200 Hz sampling frequency, with 2-4 repeated runs at low speeds (Fr = 0.769 & Fr = 0.961) and 3-5 repeated runs at high speeds (Fr =1.153 & Fr = 1.354). The data acquisition time was around 6 s to 9 s and 2.5 s to 4.3s for low and high speeds respectively.

The uncertainty analysis was carried out following the ITTC guidelines 7.5-02-02-02 and 7.5-02-02-02.1 [17], [18]. Examples of the uncertainty of resistance, trim and sinkage are given in Appendix A.

The bias of the force gauge (dynamometer) was 0.2% in the range of 0 to 0.2 kN. The dynamometer was calibrated before the model test according to the ITTC Procedure 7.5-01-03-01 [19]. Five loads were implemented by weights and randomly applied three times each load. The fitting curve for predicting the force was obtained by linear regression with a standard deviation of SEE = 0.077N.

The trim of the craft was evaluated from the readings of draught change in the aft and the bow. The draught changes were measured with linear measuring wire and draw-wire encoder the bias of which is 0.05% in a range of ±1250mm.

The tachometer was calibrated at 1 m/s, 2 m/s, 3 m/s, 4, m/s and 5 m/s, four runs for each speed except for the speed 1m/s which was run 6 times. The fitting curve for predicting the speed was obtained by linear regression with a standard deviation of SEE=0.0011m/s.
The temperature of the water in the tank was measured once a day with a K-type thermo-couple the bias of which is 1 °C in a range of -40 °C to 1372 °.

3. RESULTS AND DISCUSSION

The two tested sets of chine modifications were compared to their respective benchmarks in terms of total resistance, trim and sinkage at the *CG*. Figures 4-6 show the comparison in the design condition of the craft (Δ_m = 40 kg and Fr =1.354). In Figures 7-9, the test results are presented as functions of Froude number, where the Fr is based on the craft's length between perpendiculars L_{PP} = 1.8 m for better visualization.

Compared to the bare hull (BH), at the design speed of Fr = 1.345, the chine spray interceptor (BH-CSI) and chine inclination (BH-C.INC) reduced the total resistance by 2.6% to 3.4% and 1.4% to 2.9% respectively. The change of resistance was also dependent on the displacement. The lighter the craft the greater the reduction of total resistance.
The influence of the chine spray interceptors on the performance of the boat equipped with spray rails was rather similar to that of the bare hull. The chine spray interceptors increased the total resistance by around 1% in semi-displacement condition and decreased it by up to 3% in planing condition.

In semi-displacement regime, the rectangular 2x2 mm chine spray interceptor (SR-CSI) increased the total resistance in light and heavy loading conditions respectively by 1.5% - 3.2% and 0.8% - 2%. In planing regime the total resistance was increased by 0.4% - 1.1% in light loading conditions and it was decreased by 0.8% - 1.3% in heavy loading conditions.

In the semi-displacement regime, the triangular 5x5 mm chine spray rail (SR-CSR) increased the total resistance by up to 0.9% and 1.9% in light and heavy loading conditions respectively. In planing regime, the total resistance decreased by 2.1% - 2.6% and 2.5% -3% in light and heavy loading conditions respectively.

Similarly, to the cases with the bare hull, both the chine spray rail and chine interceptor increased the running trim in semi-displacement and reduced it in planing mode. The chine spray interceptors also reduced the sinkage at the CG through the whole speed and displacement range. However, the increase of the sinkage at the CG was less than that in the case of the bare hull.

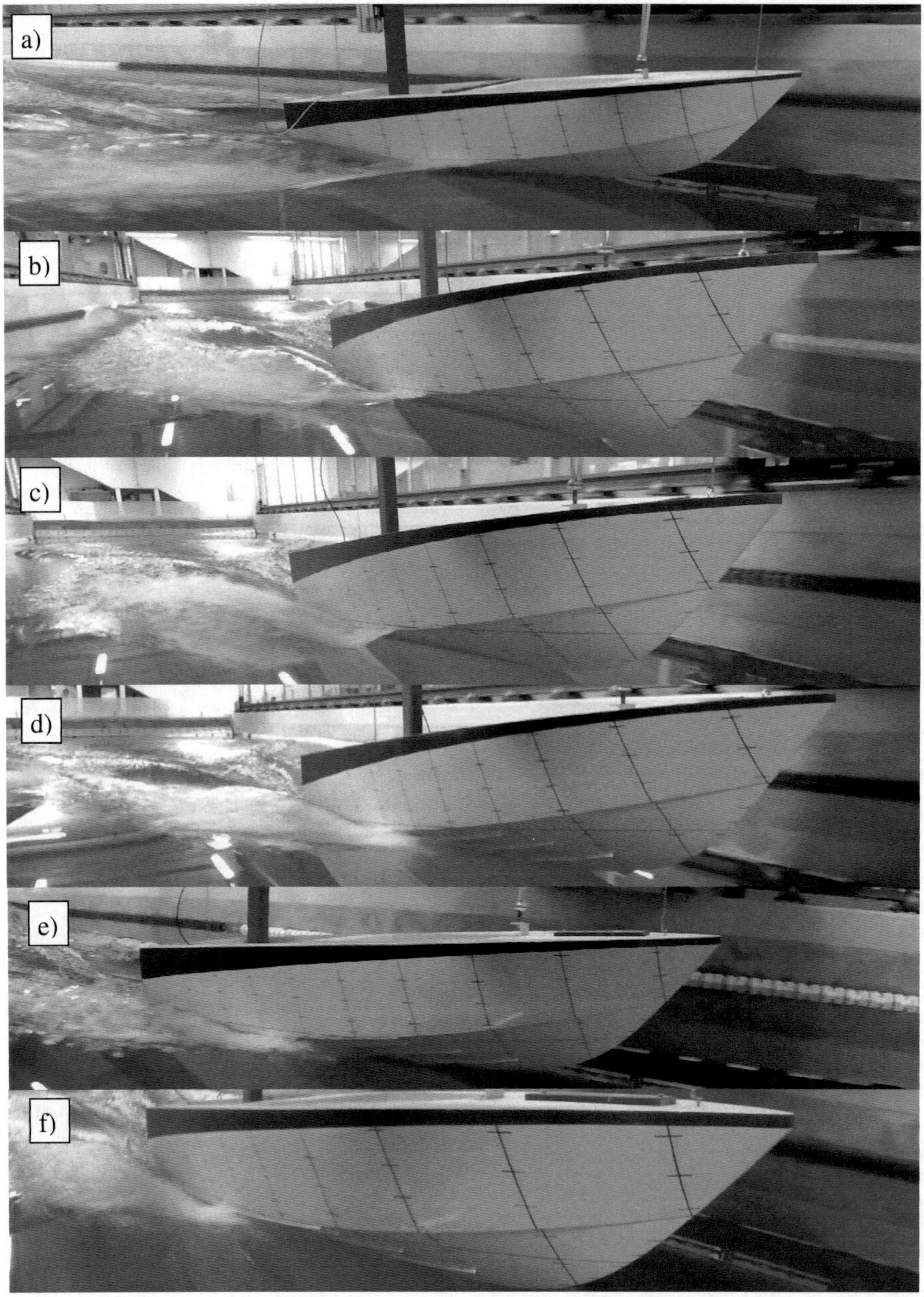

Figure 3: Spray formation on the tested models in design condition Fr = 1.354 and Δm = 40kg: a) BH, b) BH-C.INC, C)BH- CSI, d) SR, e) SR-CSI, f) SR-CSR.

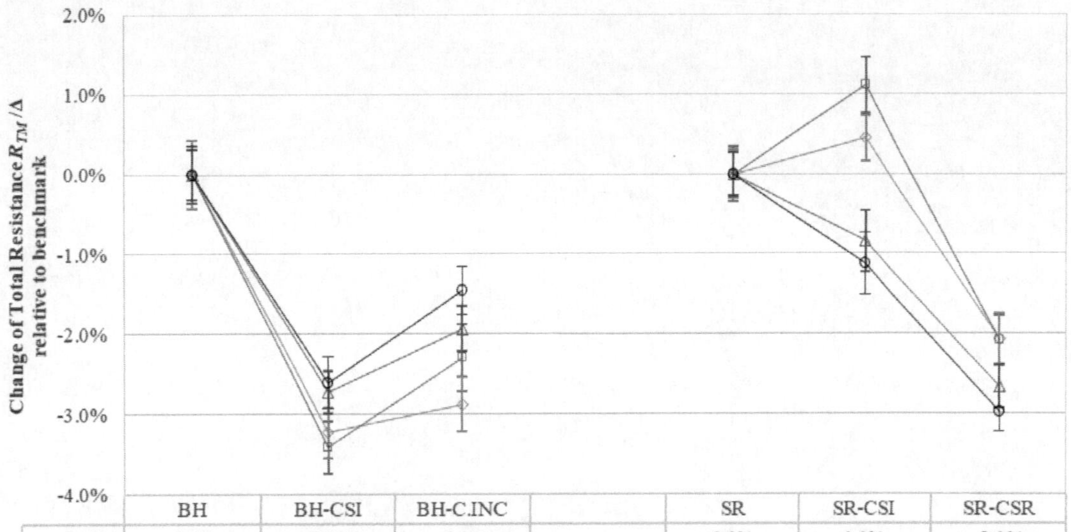

	BH	BH-CSI	BH-C.INC		SR	SR-CSI	SR-CSR
30kg	0.0%	-3.4%	-2.3%		0.0%	1.1%	-2.1%
35kg	0.0%	-3.2%	-2.9%		0.0%	0.4%	-2.1%
40kg	0.0%	-2.7%	-1.9%		0.0%	-0.8%	-2.7%
45kg	0.0%	-2.6%	-1.4%		0.0%	-1.1%	-3.0%

Figure 4: Total resistance R_{TM}/Δ with expanded uncertainty $U'_p\ (R_{TM}/\Delta)$ of the tested configurations relative to their benchmarks in design condition $Fr = 1.354$ and $\Delta_m = 40$kg.

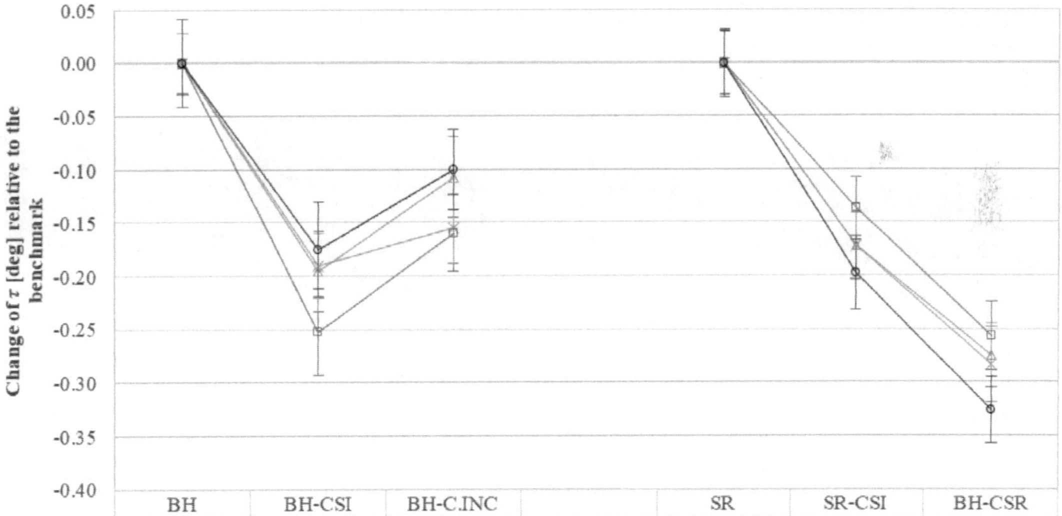

	BH	BH-CSI	BH-C.INC		SR	SR-CSI	BH-CSR
30kg	0.00	-0.25	-0.16		0.00	-0.14	-0.26
35kg	0.00	-0.20	-0.11		0.00	-0.17	-0.27
40kg	0.00	-0.19	-0.16		0.00	-0.17	-0.28
45kg	0.00	-0.18	-0.10		0.00	-0.20	-0.33

Figure 5: Dynamic trim τ with expanded uncertainty $U_p\ (\tau)$ of the tested configurations relative to the bare hull in design condition $Fr = 01.354$ and $\Delta_m = 40$kg.

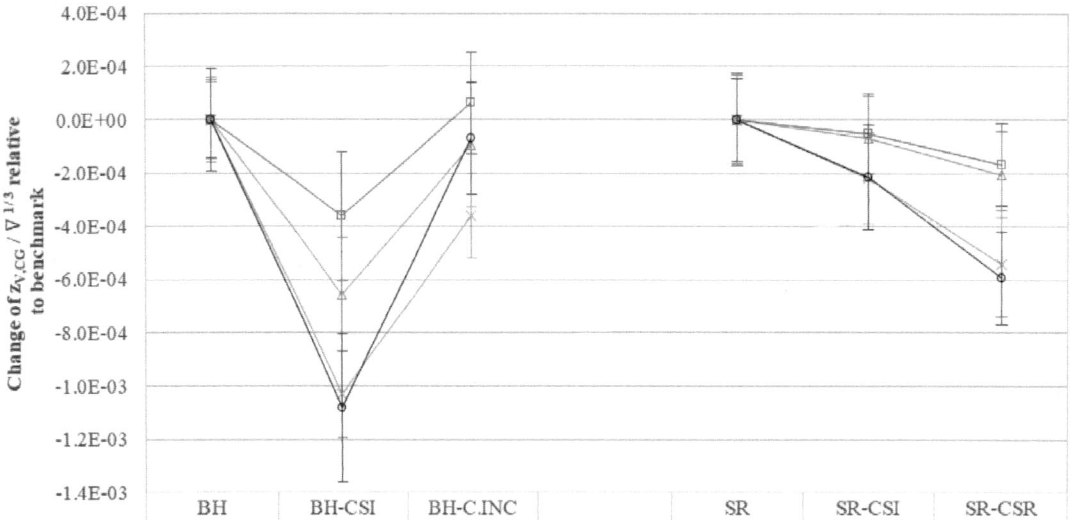

	BH	BH-CSI	BH-C.INC		SR	SR-CSI	SR-CSR
Series1	0.0E+00	-3.6E-04	6.3E-05		0.0E+00	-5.2E-05	-1.7E-04
Series2	0.0E+00	-6.6E-04	-9.5E-05		0.0E+00	-6.9E-05	-2.1E-04
Series3	0.0E+00	-1.0E-03	-3.6E-04		0.0E+00	-2.2E-04	-5.4E-04
Series4	0.0E+00	-1.1E-03	-6.9E-05		0.0E+00	-2.2E-04	-5.9E-04

Figure 6: Non-dimensional dynamic sinkage at the centre of gravity $z_{V,CG} / \nabla^{1/3}$ with expanded uncertainty U_p ($z_{V,CG} / \nabla^{1/3}$) of the tested configurations relative to the bare hull in design condition Fr = 01.354 and Δ_m = 40kg.

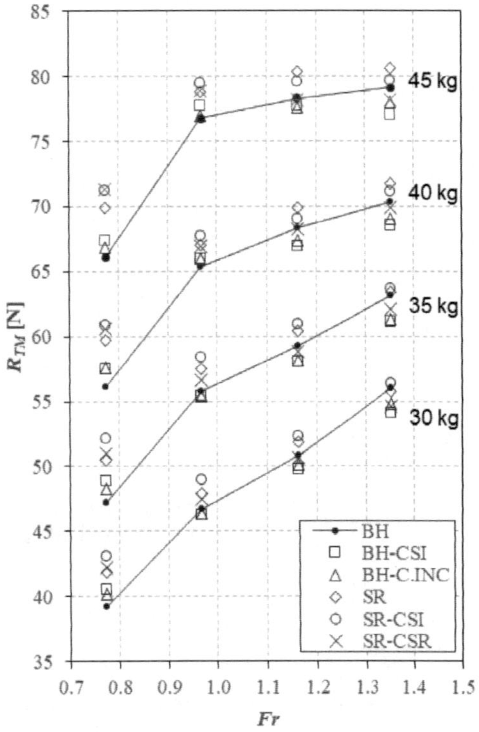

Figure 7: Total resistance R_{TM} of the tested configurations.

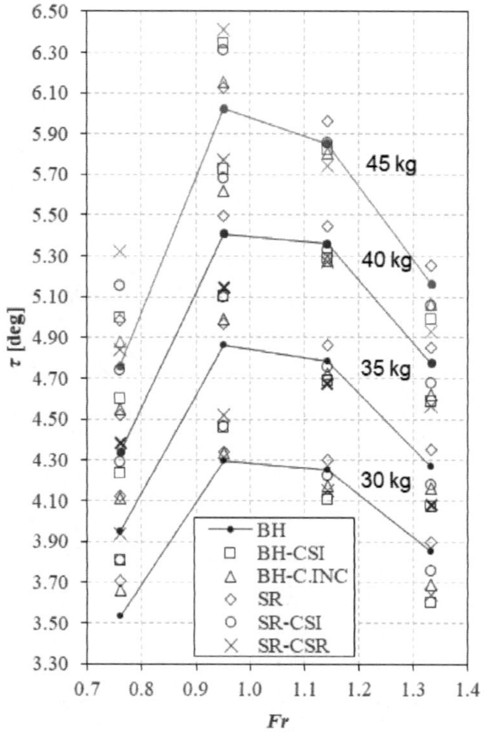

Figure 8: Dynamic trim τ of the tested configurations.

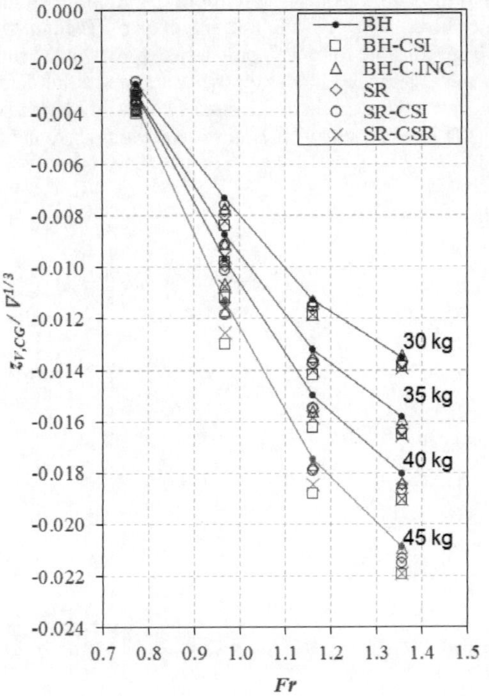

Figure 9. Non-dimensional dynamic sinkage $z_{V,\,CG} / \nabla^{1/3}$ at the centre of gravity of the tested configurations.

4. CONCLUSIONS

The results of an experimental series comprising two benchmarks and four chine modifications are presented in this paper. Based on the results following conclusions can be made:

- The results show that it is possible to affect the running position of planing craft with rather simple chine modifications.
- Both chine spray interceptor and chine inclination reduced trim and total resistance of the craft, particularly at high speed and in heavy loading condition.
- A simple rectangular interceptor on the chine reduced the total resistance and trim nearly twice as much as the chine inclination did.
- A triangular chine spray rail reduces the resistance and running trim more than a rectangular one does.

The main problem of this paper is that it remains unclear whether the triangular 5x5mm chine spray rail performed better than rectangular 2x2mm chine interceptor, due to its size or angle with the chine surface. Therefore, future studies should include a larger variety of tested chine modifications to separate the influence of the angle of the spray deflection surface from that of its size.

5. ACKNOWLEDGEMENTS

This research work was financed by the Estonian Research Council via grant PRG83 (Numerical simulation of the FSI for the dynamic loads and response of ships) and by the European Regional Development Fund via grant 2014-2020.5.04.19-0385 (Small Craft Competence Centre – analysis of the impact of external construction elements on the performance of a vessel, transfer of know-how and information). This help is gratefully appreciated

6. REFERENCES

[1] E. P. Clement, "Effects of Longitudinal Bottom Spray Strips on Planing Boat Resistance, Report No.1818," Washington, DC., 1964.

[2] E. P. Clement, "Reduction of Planing Boat Resistance by Deflection of the Whisker Spray, Report No.1929," Washington, DC., 1964.

[3] G. Grigoropoulos and T. Loukakis, "Effect of Spray Rails on the Resistance of Planing Hulls," in *3rd Intl. Symposium on Fast Sea Transportation FAST'95*, 1995, no. September.

[4] G. Grigoropoulos, "The use of Spray Rails and Wedges in Fast Monohulls," in *IV High-Speed Vehicles Intl. Conf. HSMV'97*, 1997, no. March.

[5] W. J. Kapryan and G. M. Boyd, "The Effect of Vertical Chine Strips on the Planing Characteristics of V- Shaped Prismatic Surfaces Having Angles of Dead Rise of 20 Degrees and 40 Degrees," Langley Field, VA, 1953.

[6] D. Savitsky, M. F. DeLorme, and R. Datla, "Inclusion of Whisker Spray Drag in Performance Prediction Method for High-Speed Planing Hulls," *Mar. Technol.*, vol. 44, no. 1, pp. 35–56, 2007.

[7] D. Savitsky, "Hydrodynamic design of planing hulls," *Marine Technology*, vol. 1, no. 1. pp. 71–94, 1964.

[8] D. Savitsky and J. P. Breslin, "On the Main Spray Generated by Planing Surfaces," Hoboken, New Jersey, 1958.

[9] D. Savitsky and M. Morabito, "Origin and Characteristics of the Spray Patterns Generated by Planing Hulls," *J. Sh. Prod. Des.*, vol. 27, no. 2, pp. 63–83, 2011.

[10] T. C. Fu *et al.*, "An Assessment of Computational Fluid Dynamics Predictions of the Hydrodynamics of High-Speed Planing Craft in Calm Water and Waves," in *30th Symposium on Naval Hydrodynamics*, 2014, no. November.

[11] J. Seo *et al.*, "Model tests on resistance and seakeeping performance of wave-piercing high-speed vessel with spray rails," *Int. J. Nav. Archit. Ocean Eng.*, vol. 8, no. 5, pp. 442–455, 2016.

[12] D. J. Kim, S. Y. Kim, Y. J. You, K. P. Rhee, S. H. Kim, and Y. G. Kim, "Design of high-speed planing hulls for the improvement of resistance and seakeeping performance," *Int. J. Nav. Archit.*

Ocean Eng., vol. 5, no. 1, pp. 161–177, 2013.
[13] L. Olin, M. Altimira, J. Danielsson, and A. Rosén, "Numerical modelling of spray sheet deflection on planing hulls," in *Proceedings of the Institution of Mechanical Engineers, Part M: Journal of Engineering for the Maritime Environment*, 2016, vol. 231, no. 4, pp. 811–817.
[14] B. Molchanov, S. Lundmark, M. Fürth, and M. Green, "Experimental validation of spray deflectors for high speed craft," *Ocean Eng.*, vol. 191, no. April, p. 106482, 2019.
[15] M. Lakatos, T. Sahk, R. Kaarma, K. Tabri, M. Kõrgesaar, and H. Andreasson, "Experimental testing of spray rails for the resistance reduction of planing crafts," in *Trends in the Analysis and Design of Marine Structures : Proceedings of the 7th International Conference on Marine Structures, MARSTRUCT 2019*, 2019, pp. 334–343.
[16] 28th ITTC, "7.5-02-02-01 Resistance Test," ITTC, 2017.
[17] 27th ITTC, "7.5-02-02-02 General Guideline for Uncertainty Analysis in Resistance Tests," 2014.
[18] 27th ITTC, "7.5-02-02-02.1 Example for Uncertainty Analysis of Resistance Tests in Towing Tank," 2014.
[19] 28th ITTC, "7.5-01-03-01 Uncertainty Analysis, Instrument Calibration," 2017.

7. AUTHORS BIOGRAPHY

Mikloš Lakatoš holds the current position of early-stage researcher at Tallinn University of Technology, Small Craft Competence Centre. He is responsible for on experimental testing and numerical modelling of spray rails for improved resistance and seakeeping performance of the fast medium-sized craft. His previous experience includes work as a naval architect specialising in marine hydrodynamics, hull design and propulsion calculations in at Deltamarin Ltd.

Tarmo Sahk holds the current position of head of laboratories at Small Craft Competence Centre of Tallinn University of Technology. He is responsible for conducting model testing and laboratory services for commercial and research purposes. His previous experience includes a BSc in Small Craft Building.

Henrik Andreasson holds the current position as Naval Architect and owner of Flow Naval Architects. He has been working with hydrodynamic design and optimization since 1999. Until 2009 he worked as project manager at the model test facility SSPA Sweden AB. Since 2015 he has been Hydrodynamic mentor at Tallinn University of technology, Small Craft Competence Center, involved within the development and running of the towing tank.

Kristjan Tabri holds the current position of senior research scientist at Tallinn University of Technology. He is responsible for developing a two-way coupled fluid structure interaction model for dynamic impact problems. His previous experience includes the simulation of structural behaviour of marine structures under dynamic loads.

EVALUATION OF CREW TRANSFER VESSEL PERFORMANCE: NUMERICAL SIMULATIONS, ONBOARD MEASUREMENTS AND MODEL TESTS

J. W. Serraris, G. D. Struijk, E. F. G. van Daalen, F. Sterlini, S. Gueydon, F. Miguel Montero, K. M. Jansen, Maritime Research Institute Netherlands (MARIN), The Netherlands
J. W. van Heemst, E. Hoogerwerf, BMO Offshore, The Netherlands

SUMMARY

Crew Transfer Vessels (CTVs) sail between local ports and offshore wind farms to transport maintenance technicians. To maximise wind turbine uptime weather dependent planning of maintenance trips is an important task. In this research, numerical simulations are performed to predict vessel motions and seasickness during transit trips. The numerical simulations are compared to onboard measurements and model tests. The comparison between onboard measurements and numerical predictions shows significant discrepancies. These differences originate from uncertainties in vessel properties and environmental conditions used as input for the numerical simulations. The comparison between model tests in controlled conditions and numerical simulations shows a much better agreement. This isolates the uncertainties in the numerical model. To assist in daily planning, it is recommended to increase the accuracy of input parameters to improve the quality of numerical motion and seasickness predictions. Futhermore it is recommended to apply onboard mitigation methods to increase comfort.

1 INTRODUCTION

While more and more offshore wind farms are installed and operated, maintenance on the wind turbines is required to maximise the energy yield. Maintenance work comes in large variety, from replacements to repairs. Each maintenance operation requires its own specialist and specialised vessel.

For both proactive maintenance (e.g. regular service and upgrades to turbine components), and reactive maintenance (e.g. response to alarms and failures) of offshore wind farms at close range to the shore, Crew Transfer Vessels (CTVs) are used to transport technicians to and from the wind turbines on a daily basis. For this catamarans with 20 - 30 m length are typically used. These small craft operate at high speeds of about 20 - 25 knots and often in moderate sea states. This operational profile results in relatively large ship motions and accelerations. This poses threats to the wellbeing, safety and productivity of the technicians.

To maximise wind turbine uptime weather dependent planning of maintenance trips to the offshore locations is an important task. Historically, planning depended on rules-of-thumb for working limits on vessels. Decisions on whether or not to sail are taken in advance, based on a threshold for wave height, predicted vessel motions and the local experience of the marine coordinators. Thresholds for CTV operations are given in Ref. [1] and [2]. If sailing occurs, then the willingness of the vessel captain and technicians to perform a safe transfer to a turbine depends on their experience during the transit and while at site. This approach has a significant opportunity cost: on some days operations are not attempted, when they could have been conducted; other days operations are attempted when they cannot be safely conducted, particularly due to sea-sickness.

1.1 THE SPOWTT PROJECT

The safety and productivity of offshore wind farm technicians can be improved by better predicting the physiological and psychological effects of transport in a CTV. This challenge led to the formation of the multi-disciplinary and multi-national project SPOWTT: "improving the Safety and Productivity of Offshore Wind Technician Transit". The main objective was to improve the health and safety of technicians by developing an evidence-based, open access model to provide more effective advice for the 'sail/no sail' decision. To enable this, the modelling approach as depicted in Figure 1 was taken. Research efforts focussed on increasing knowledge about both the CTV motions and the human reaction to those motions. The overall SPOWTT methodology is described in Ref. [3].

This paper describes the efforts made and lessons learned within SPOWTT on the motion behaviour of CTVs. Investigations have been made using numerical simulations, model tests and full scale performance data from an elaborate measurement campaign on different CTVs over multiple years' time.

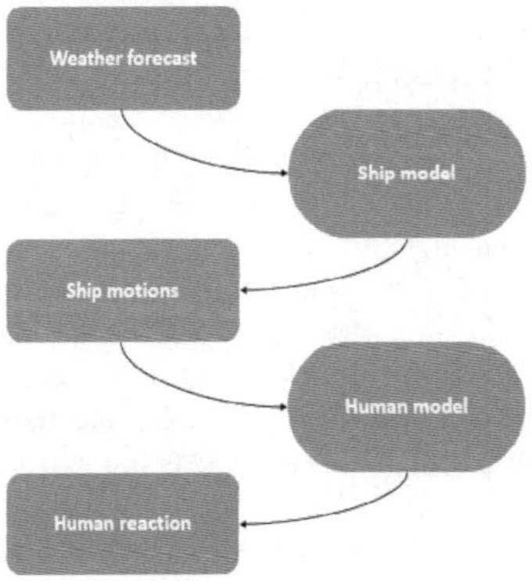

Figure 1: SPOWTT model

2 OBJECTIVES AND APPROACH

The objective of this research is to obtain trustworthy predicted motions and accelerations of CTVs during transit trips. These serve as the basis for the prediction of seasickness and will be used by the onshore marine coordinators for daily planning. To support the decision making, the predictions have to be available prior to the operation.

Onboard measurements are the most accurate representation of the in-situ motions and accelerations. However, these are available only during the transit trip and are therefore not suitable to support the planning phase prior to operations.

In order to be available prior to operations the motions and accelerations are calculated by numerical simulations. Onboard measurements and model tests are performed to validate the results of the numerical predictions. This relation is illustrated in Figure 2.

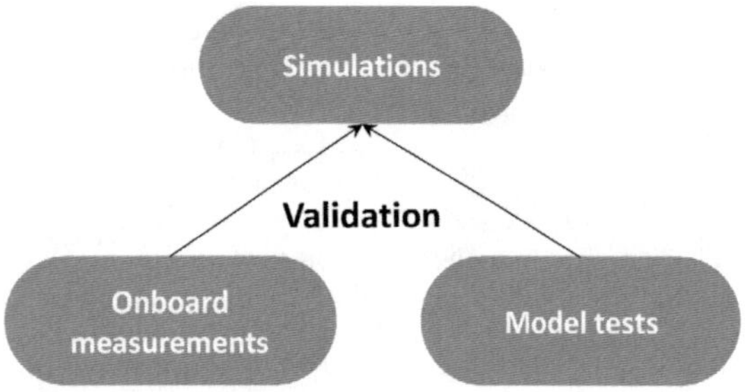

Figure 2: Schematic relation between Simulations, Onboard measurements and Model tests

Onboard measurements are the most realistic source for validation. However, these contain uncertainties on:

- Properties of the vessel: the actual hull geometry, trim flap angle and weight distribution are not known exactly. These are taken from public factsheets and onboard stability books.
- Environmental conditions: the actual seastates encountered during the voyages are not known. Only spectral weather forecasts and hindcast are available.
- Unsteadiness of the course, speed and environmental conditions during of the transit trips. The numerical simulations represent steady conditions. For the validation transit trips with minimal variations in the conditions are selected. Despite this, actual conditions always show some unsteadiness.
- Calculation method: the applied existing hydrodynamic software PANSHIP is validated for various types of vessels, but not specifically for CTVs. To isolate and address the uncertainty of the calculation method, model tests are performed with known vessel particulars and known environmental conditions. The model tests are described in Section 5.

3 NUMERICAL SIMULATIONS

Numerical predictions can be performed "live" based on the environmental forecast, or a priori and stored in a database, which is used as a lookup table once the environmental forecast is available. The two approaches are described in Ref. [4]. The direct calculation approach provides the most accurate prediction based on the actual weather forecast, but requires dedicated numerical software and expertise in handling this software. The database approach is more robust and faster, but requires many entries to cover a range of vessel properties, vessel speeds and potential environmental conditions. Because of the robustness and the intended use of the results in a practical planning tool the database approach is followed in the present research project.

The databases containing the vessel motion response and human seasickness are incorporated in an onshore decision support tool developed by SPOWTT partners TNO and SMC, see Ref. [3].

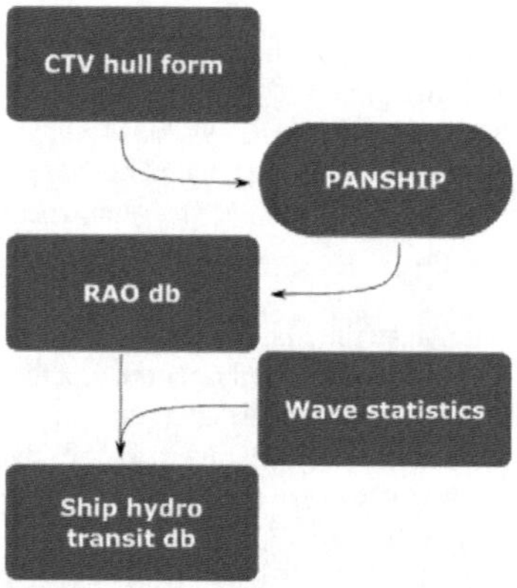

Figure 3: schematic numerical approach

3.1 NUMERICAL APPROACH

The simulations are performed with MARIN's existing non-linear hydrodynamic simulation code PANSHIP, version IRF 3.1.0. This numerical code uses impulse response functions (IRF) to compute the vessel motions. The code includes disturbed wave effects, linear radiation and diffraction forces, nonlinear excitation and restoring forces. The code and its applications are described in Ref. [5-6]. The software allows for non-linear effects, such as lifting and slamming, into account. In order to apply the results to a wide range of seastates, the hydrodynamic simulations are linearised. This means that each sea state is defined as a linear sum of sine components of different frequencies. The results of the linearised calculations are stored as Response Amplitude Operators (RAOs) in an "RAO db".

The vessel motion calculations are performed for the following variations:

- Six realistic catamaran type CTVs ranging from 13m through 25m. The hull geometry (lines plan), main particulars (loading conditions and stability data) and appendages data (trim flaps) are obtained from public factsheets, general arrangement (GA) drawings and on-board stability books. The main particulars of the CTVs are listed in Table 1.
- Three speeds: 15, 20 and 25 knots. The typical sailing speed of a CTV is 25 knots. The lower speeds allows to account for the positive effects of voluntary speed losses on the motion behaviour and thus on the sea-sickness; and thereby include the effects in the planning.
- Approximately 414,000 sea states consisting of oblique wind sea and swell waves. Wave conditions consists of variations in wave height (0.25m to 2.00 m with step 0.25m), wave period (2s to 9 s with step 1s) and wave directions (0° to 360° with step 30°). A JONSWAP wave spectrum (γ=3.3) and the cos-2s wave spreading model (with s=10) was applied to represent realistic conditions.

For each combination of CTV, speed and wave condition the vessel motions and accelerations are calculated and represented by the Root Mean Squared (RMS) as calculated by Eq. [1-2].

$$S_x(\omega_0,\theta)=S_\xi(\omega_0,\theta)H^2(\omega_0,\theta) \qquad \text{Eq. [1]}$$

$$RMS_x=\int_0^\infty\int_0^{2\pi} S_x(\omega_0,\theta)\,\mathrm{d}\theta\,\mathrm{d}\omega_0 \qquad \text{Eq. [2]}$$

where:

ω_0	Wave frequency (rad/s)
θ	Wave direction (deg)
H	Response Amplitude Operator (m/m)
$S\xi$	Wave energy spectrum density (m^2 s)
S_x	Response spectrum density (m^2 s)

The seasickness is calculated by the SPOWTT seasickness formula describes in Ref [3]. The results of these motion calculations are stored in tabular format such that they can be read and used by O&M planning and decision support tools.

4 ONBOARD MEASUREMENTS

4.1 PARTICULARS AND CONDITIONS

Throughout the SPOWTT project, the BMO Vessel Motion Monitoring System (VMMS), Ref. [10], has gathered data on all subject vessels. Measurements have been taken on 14 different vessels from 5 distinct operators across 5 wind fields. These vessels ranged from small 40 tons CTVs with a relatively short hull length (13m) towards large 80 tons vessels of 27 m length. Hull geometries are derived from available General Arrangements (GA). The particulars are obtained from public information sheets and onboard stability books. A summary is presented in Table 1.

In total 2,071 days of operation have been monitored. It is estimated that this has resulted in ~2,500 transit trips between port and wind farm.

In the remainder of this section we outline the parameters recorded, the parameters computed, and the derived statistics used in the other sections.

4.2 RECORDED AND DERIVED PARAMETERS

4.2(a) Recorded Parameters

The VMMS records the following parameters:

- GPS location of the vessel (latitude, longitude)
- Absolute speed of the vessel
- Heading of the vessel
- Translational accelerations in 3D
- Angular accelerations around the 3D axes
- Vessel attitude (pitch, roll, yaw)

The former three parameters (location, speed and heading) have been recorded at a frequency of 1 Hz. The latter parameters (accelerations and attitude) were recorded at 40 Hz.

4.2(b) Derived Parameters

The goal of this research is to reduce motion induced sickness. This is a very subjective phenomenon and there is no single consensus on the definition of how motion sickness should be computed. Hence, the following sickness indicators are calculated:

- Motion Sickness Indicator (MSI), Ref. [8].
- Vibration Dose Value (VDV), Ref. [8].
- Illness Rating (IR). The Illness Rating is developed by TNO, MARIN, the Dutch Navy and Feadship and is an enhanced version of MSI. It includes 3 DOF accelerations, susceptiability and habituation
- A newly developed SPOWTT seasickness formula, Ref. [1].

The recorded accelerations were converted to power spectral densities (PSD) via the Fast Fourier Transform (FFT) method, Ref. [9]. PSDs are used to compare the recorded motions in the time domain to the predicted motions from simulations in the spectral domain.

4.3 VALIDATION OF SIMULATIONS

4.3(a) Approach

The numerical simulations are performed in the time domain. The results are translated to the frequency domain as described in Section 3.1. For validation of the numerical results the signals measured during transit trips are converted to the frequency domain as well. The spectral analysis requires trips with steady conditions: limited variations in vessel course, speed and environmental conditions and a duration of minimum 20 minutes. This resulted in 112 voyages with steady conditions.

In addition, the environmental conditions during the transit trips are required for the validation analysis. The wave conditions during the transit trips are collected from the Copernicus 027 hindcast model, Ref. [7]. This hindcast model provides the peak wave direction μ, the significant wave height H_s and the wave peak period T_p for wind waves, swell and the combined wave field. The model enables data in high spatial resolution (1/12°x1/12°) with 30 minute intervals. The model lacks information about the wave spectrum type and the wave directional spreading. In this study, the JONSWAP wave spectrum and wave spreading according to the cos-2s model are assumed with s defined as a function of H_s and T_p.

To validate the simulation results, the PSD of the motions and accelerations recorded onboard are compared with those predicted by the simulations. This is a valid approach, as all four sickness indicators can be computed based on the acceleration PSDs. Hence, if comparable results for the motion and acceleration PSDs are obtained, it is expected that the sickness predictions are comparable as well. This is preferred to comparing the resulting sickness ratings directly, because the motions give us a better understanding of the overall performance of the models. The focus for this comparison is between 0.1Hz and 0.5Hz, since all four seasickness models are focussed on motions in that interval.
The PSDs of the actual motions are calculated onboard by BMO. The PSDs of the numerical simulations are calculated using Eq. [1].

4.3(b) Conclusions

A comparison is made between the results of the numerical simulations and 112 steady transit trips. An example of such an analysis and comparison is shown in Figure 6. In general, the results of the numerical calculations and onboard measurements show significant differences. These differences originate from:

- Uncertainties in the properties of the actual CTVs.
- Uncertainties in the actual environmental conditions.
- Unsteadiness of the course, speed and environmental conditions during of the transit trips.
- Unidentified non-linear contributions: the on-board measurements show low frequent and wave frequent response. The linearised calculations result in wave frequent response only. On the other hand, low frequent motions have marginal influence on sea-sickness.
- Uncertainties in the numerical method. To isolate and address the uncertainty of the calculation method, model tests are performed with known vessel particulars and known environmental conditions. The model tests are described in Section 5.

5 MODEL TESTS

5.1 CTV PARTICULARS

The model tests are performed with a generic CTV design. The design is made by MARIN and represents a catamaran type CTV of 25m length and 90 tons displacement. The hull geometry is shown in Figure 4 and an impression of the model is shown by Photo 1. The model is equipped with two water jet units and an additional detachable T-foil. The T-foil was included in part of the test programme to assess its stabilising effect and consequent benefits on comfort and operability. In the remainder of this publication only the tests without the T-foil are described, since this configuration agrees with the full scale measurements and numerical simulations.

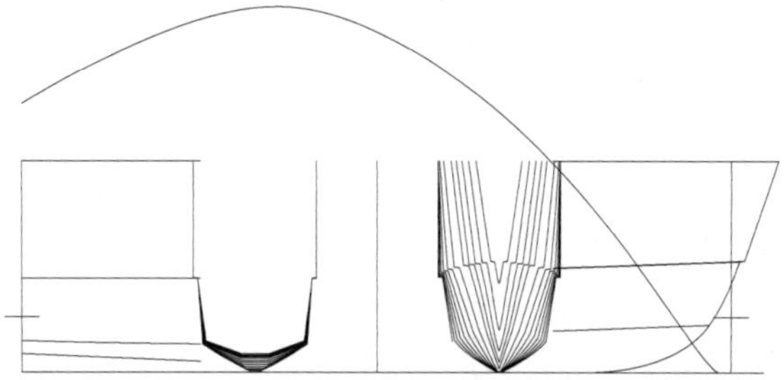

Figure 4: Lines plan of MARIN generic CTV

Photo 1: CTV model used for model tests

5.2 TEST CONDITIONS

The model tests were performed in MARIN's Seakeeping and Manoeuvring Basin (SMB). The basin dimensions are 170m length x 40m width x 5m depth. It is equipped with wave makers along one long side and one short side.

The tests were performed for the following conditions:

- CTV with and without T-foil;
- Vessel speeds 20kn and 25kn;
- Wave directions: head on (180°), bow quartering (135°), beam on (90°), stern quartering (315°) and stern on (0°);
- Regular waves ranging from wave period (T) 3.5s to 6.0s with step 0.5s and wave amplidue (H_a) of

0.375m. These conditions are tested to validate the RAOs from the numerical simulations;

- Longcrested irregular waves with wave period (T_P) of 5.0s, 6.5s and 8.0s and significant wave height (H_s) of 0.75m and 1.5 m. These conditions are tested to validate the PSDs from the simulations for irregular seastates.
- Shortcrested irruglar waves with wave period (T_P) of 5.0s and wave height (H_s) of 0.75m and 1.5m. These conditions are tested to validate the PSDs from the simulations for realistic seastates including directional spreading and to assess the effect of directional spreading on the motion response.

In addition, motion decay tests in calm water at zero speed, 20kn and 25kn were performed to obtain heave, roll and pitch damping.

5.3 RECORDED AND DERIVED PARAMETERS

Model test measurements included: incoming wave height, 6 Degree Of Freedom (DOF) motions and accelerations, relative wave heights along the hull, waterjets rpm, steering angle, inlet and outlet pressure and foil flap angle. Measurements of incoming wave height, 6 DOF motions and rpm were performed at 100Hz. Accelearations and all other parameters were measured at 200Hz. The measured time traces were applied to derive statistics, PSDs and RAOs.

5.4 VALIDATION OF SIMULATIONS

From the tests in regular waves RAOs are derived by harmonic analysis and compared to the RAOs from the numerical simulation. This comparison is shown in Figure 5 and shows a good agreement.

For the tests in irregular waves a statistical comparison is performed based on the ratio of the standard deviations of the motion or acceleration component and the wave elevation (Stdev. signal / Stdev. ζ). This approach cancels out the effect of wave height. The results of this comparison are shown in Figure 7. The results of the tests in shortcrested waves are included in this figure too.

The following is observed and concluded from the results presented in Figure 5 and Figure 7:

- In general, simulated results are close to model tests results. The RAOs derived from the tests in regular waves match very well. The standard deviation ratio derived from the tests in longcrested irregular waves matches good as well.
- The good agreement between the numerical simulations and model tests demonstrates that PANSHIP is able to calculate and predict CTV motions and accelerations. From this it is concluded that the large mismatch between the onboard measurements and the numerical simulations originates from uncertainties in the vessel properties and environmental conditions.
- Results of the model tests in shortcrested waves show a decrease of the response of the mode of motion in-line with the dominant wave direction eg. roll in 90° waves, and an increase in the response for modes of motion perpendiculat to the wave direction eg. pitch in 90°. This is because of the spatial spreading of the wave energy. This effect demonstrates the need to take wave directional spreading into account in the numerical predictions of CTV motions.

6 CONCLUSIONS

The present research describes a comparison between numerical simulations, on-board measurements and model tests for CTVs as part of the SPOWTT project. From this analysis the following is concluded:

- Comparison between model tests and numerical simulations shows a good agreement. This demonstrates the validaty of the numerical simulations to provide thrustworthy predictions of motions, accelerations and seasickness.
- The discrepancies between the onboard measurements and numerical simulations originate from uncertainties in the actual vessel properties and environmental conditions during the transit trips.

7 RECOMMENDATIONS

The following is recommended for future research:

- Improve the input of the numerical simulations to derive more accurate predictions of vessel motions, accelerations and seasickness. Reliable vessel properties should be obtained from the vessels designer. Trustworthy environmental conditions could be received from local met masts and/or wave buoys. For more accurate validation, onboard wave measurements, by a wave radar or ship-as-a-wave-buoy methodology, would be useful as well. The wave field parameters should include wind sea, swell and wave directional spreading.
- Onboard advice systems could assist the captain to adjust course and speed to optimise motion comfort and thereby increase the operability of CTVs and the technicians.
- Physical motion mitigation techniques, such as stabilizings fins, decrease vessel motions and thereby increase operability of ships. It is recommended to further investigate their application for CTVs.

8 ACKNOWLEDGEMENTS

The SPOWTT project was funded in part by the European Union's Horizon 2020 research and innovation programme under grant agreement No 691732. This public funding is managed in the UK by the Department for Business, Energy and Industrial Strategy (BEIS) and in the Netherlands by TKI Wind op Zee and Topsector

Energie-subsidie programme of the Dutch Ministry of Economic Affairs.

We would like to thank the following for their contribution and support: DEME Group; NR Marine Services; Seacat Services; Turner Iceni; Windcat Workboats; Galloper Wind Farm Ltd; Greater Gabbard Offshore Winds Ltd; Beatrice Offshore Windfarm Ltd.

9 REFERENCES

1. Carbon Trust, 'Crew Transfer Vessel (CTV) Performance Plot (P-Plot) Development', Research Project Summary, 2017.
2. Philips, S., Shin, I. B., Armstrong, C., 'Crew Transfer Vessel Performance Evaluation', RINA, Design & Operation of Wind Farm Support Vessels, London, UK, 2015.
3. Williams, T.M., et al, 'SPOWTT: Improving the Safety and Productivity of Offshore Wind Technician Transit', Wind Energy, 2020.
4. Serraris, J.W., Pereira, I.S., 'Workability analysis for crew transfer vessels', *3rd International Conference on Offshore Renewable Energy*, Glasgow, UK, 2018.
5. De Jong P., 'Seakeeping Behaviour of High Speed Ships, an Experimental and Numerical Study', *Ph. D. Thesis, Delft University of Technology*, 2011.
6. Van Walree F., Sgarioto D. and Turner T.G., 'Validation of a Time Domain Panel Code for Prediction of Impulsive Loads on High Speed Ships', *31th Symposium on Naval Hydrodynamics*, Monterey, California, 11-16 September, 2016.
7. Copernicus Marine Environment Monitoring Service, http://marine.copernicus.eu/
8. ISO 2631-1:1997(E) Annex D
9. Frigo, M., Johnson, S. G.. 'FFTW: An Adaptive Software Architecture for the FFT', *Proceedings of the International Conference on Acoustics, Speech, and Signal Processing. Vol. 3*, 1998, pp. 1381-1384.
10. https://www.bmo-offshore.com/Product-Services/Services

10 AUTHORS BIOGRAPHY

Ed van Daalen is Senior Researcher at MARIN's R&D Department. His experience includes frequency domain and time domain seakeeping programs, roll damping, anti-roll tanks, multi-disciplinary hull optimisation and thruster allocation.

Sebastien Gueydon was Specialist at MARIN's Offshore department until April 2020 with main expertise in numerical simulations. He has set-up the framework for the analysis of field and vessel data and their comparison with the PANSHIP simulations. His main experience for the last 10 years consists of numerical modelling and validations of offshore renewable energy devices.

Hans van Heemst is a project manager and DevOps engineer at BMO Offshore. With a background in Wind Energy and 7 years of experience in offshore wind logistics, he was responsible for coordinating succesful implementations of vessel monitoring equipment on various vessels within this project.

Erwin Hoogerwerf is data scientiest at BMO Offshore. He is responsible for data analysis and new product development. His previous experience includes his graduation project on numerical MSI forecast via machine learning.

Jorrit-Jan Serraris is Senior Project Manager at MARIN's Offshore department. He focusses on operability assessment for offshore wind turbine installation and O&M vessels. His previous experience includes operability assessments by means of model tests and numerical simulations for oil and gas platforms.

Fenneke Sterlini is specialist at MARINs Performance at Sea Department. Her current focus is on maritime decision support and data science. Her previous experience includes bridge decision support systems.

Gijs Struijk is Team Leader at MARIN's Performance at Sea department. He is focussed on performing onboard measurement campaigns to capture the real-world behaviour of ships and offshore structures. His focus is in hydrodynamics in the broad sense, with previous work ranging from seakeeping behaviour of RHIBs to full-scale cavitation observations on ship propellers.

Table 1: Main particulars of CTVs

Description	symbol	CTV-13m	CTV-14m	CTV-15m	CTV-19m	CTV-20m	MARIN generic CTV 25m
length between perpendiculars	L_{pp} [m]	12.67	14.11	14.62	19.00	19.25	25.00
length overall	L_{oa} [m]	13.86	15	17	22	22.8	N/A
mean draft	T_{mean} [m]	1.07	1.00	1.21	1.25	1.28	1.30
longitudinal centre of gravity	LCG [m]	6.18	5.45	6.48	7.78	7.25	10.19
vertical centre of gravity	KG [m]	1.86	2.17	2.17	2.59	2.50	4.28

Figure 5: RAOs of model tests and numerical simulations for regular waves

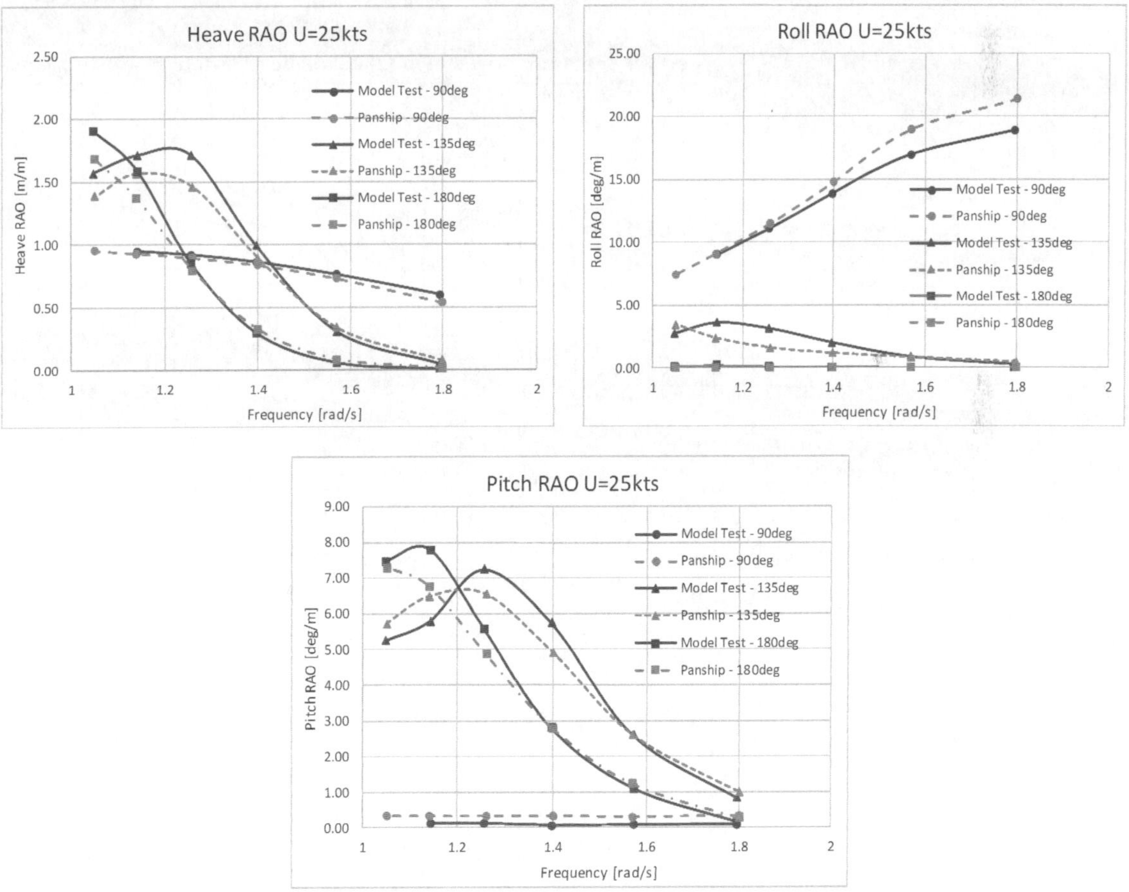

Figure 6: Transit voyage analysis

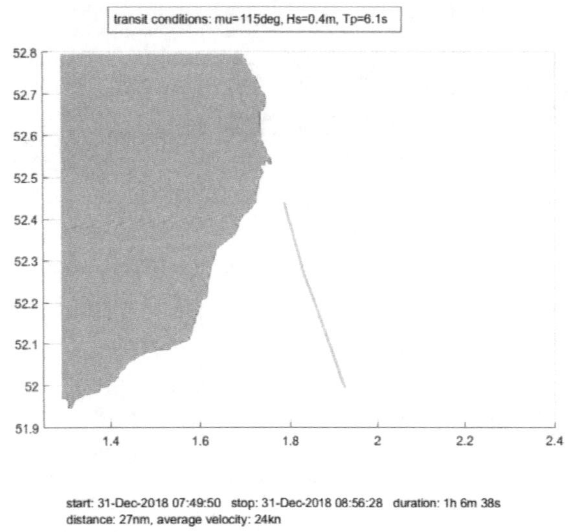

Figure 6.a: Voyage information: date, location, course and distance

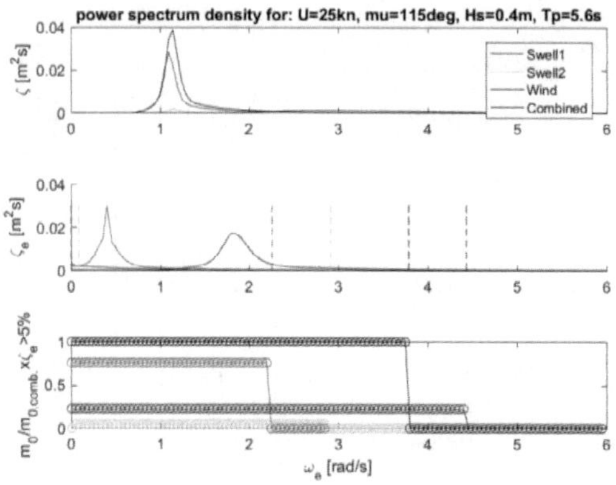

	mu_deg	Hs_m	Tp_s	m0_SI
Swell1	344	0.3	5.7	0.0071
Swell2	99	0.1	5.4	0.0004
Wind	72	0.2	2.2	0.0021
Combined	115	0.4	5.6	0.0095

Figure 6.b: Wave conditions from Copernicus

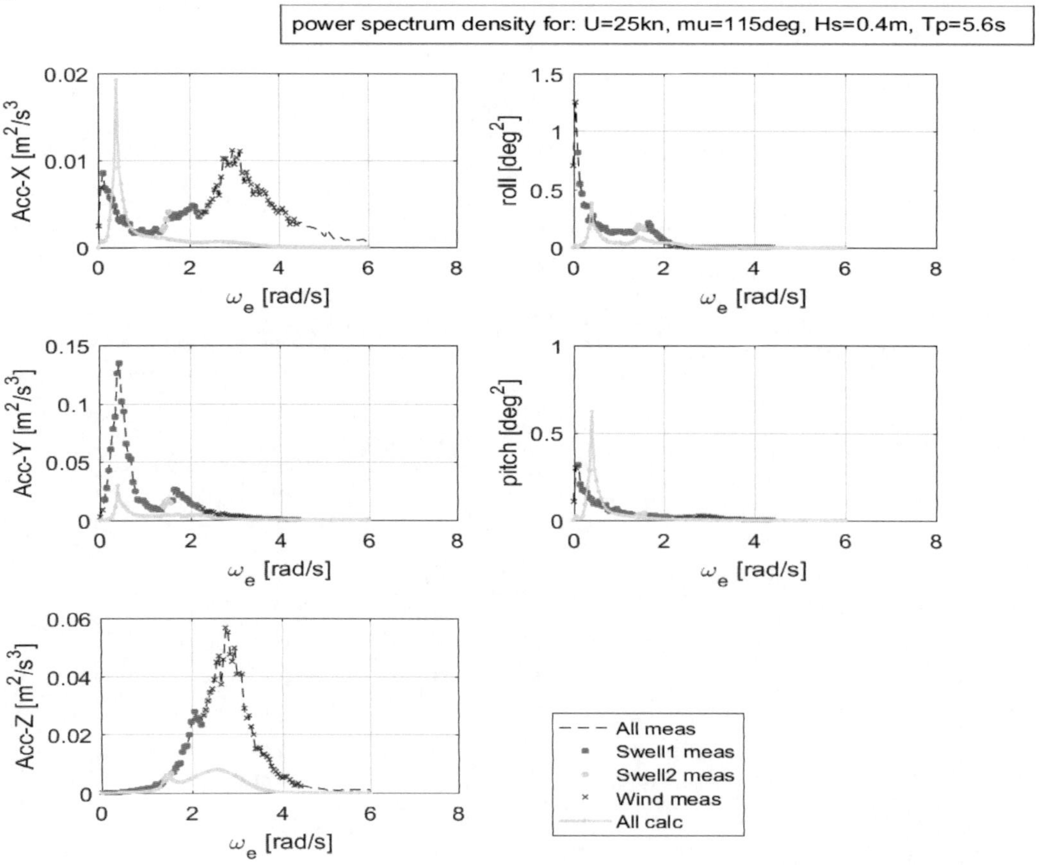

Figure 6.c: Comparison between PSD derived from onboard measurements (black dashed line with marker overlay) and calculated PSD (light blue line)

Figure 7: Resuts of model tests and numerical simulations for irregular waves H_s=0.75m and V_s=25kn

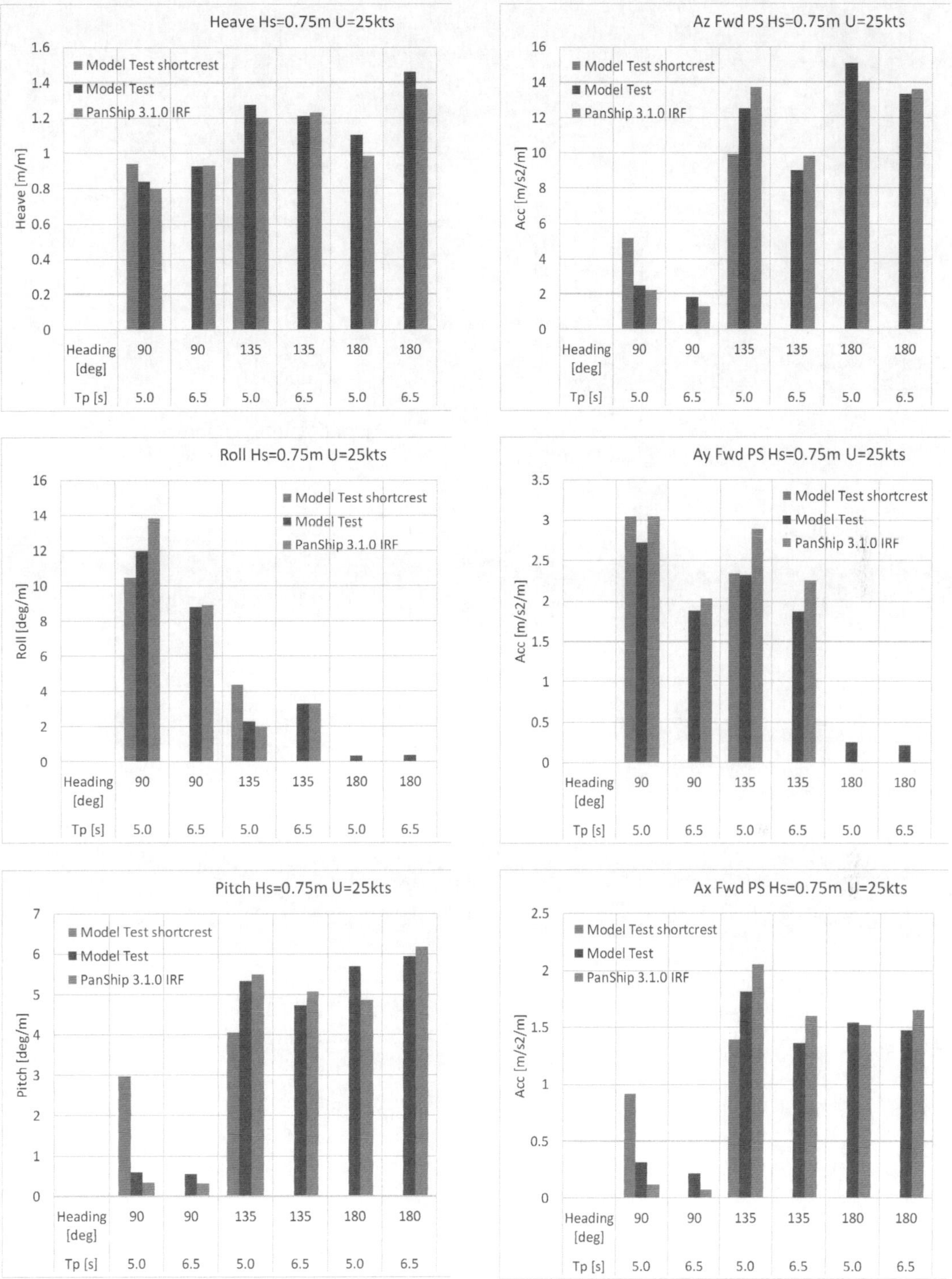